The Missouri Botanical Garden in
MADAGASCAR

Celebrating 25 Years of Exploration, Discovery, and Conservation on the "Eighth Continent"

 MISSOURI BOTANICAL GARDEN

The Missouri Botanical Garden in
MADAGASCAR

Celebrating 25 Years of Exploration, Discovery, and Conservation on the "Eighth Continent"

by Liz Fathman

with foreword by Dr. Peter Wyse Jackson

❋ MISSOURI BOTANICAL GARDEN

St. Louis, Missouri

Written and edited by Liz Fathman
Foreword by Dr. Peter Wyse Jackson
Designed by Ellen Flesch

ISBN 978-0-9884551-1-5
Library of Congress Cataloging-in-Publication Data available upon request.
1 2 3 4 5 08 09 10 11 12

Published and distributed by
Missouri Botanical Garden
Post Office Box 299
St. Louis, Missouri 63166-0299
United States
Telephone: (314) 577-5100
Website: www.mobot.org

SUSTAINABILITY

The Missouri Botanical Garden strives to make the most sustainable choices for the future of people, plants, and the planet. This book is printed on paper containing 60% post-consumer recycled content manufactured with wind power.

Environmental Benefits Statement
OF USING POST-CONSUMER WASTE FIBER VS. VIRGIN FIBER

The Missouri Botanical Garden saved the following resources by printing the pages of this book on chlorine free paper made with 60% post-consumer waste and manufactured with electricity that is offset with Green-e® certified renewable energy credits.

trees	water	energy	solid waste	greenhouse gases
22 fully grown	9,801 gallons	9 million BTUs	621 pounds	2,173 pounds CO$_2$

Calculations based on research by Environmental Defense and the Paper Task Force.

Photography by
P. Antilahimena; C. Birkinshaw; R. Bussman; L. Fathman; A. Kuhlman; A. Lehavana; P. Lowry; F. Rajaonary; F. Rakotoarivony; N. Rakotonirina; C. Rakotovao; A. Ramahefaharivelo; A. Randrianasolo; A. Rasamuel; F. Ratovoson; D. Ravelonarivo; and G. Schatz

Cover: *top left: Women prepare "pots" for seedlings at the nursery near Pointe à Larrée; middle:* Droceloncia rigidifolia, *with its delicate male flowers, occurs only in isolated dry forests of western and northern Madagascar; right: Collecting new species* Spondias tefyi *at Analavelona Sacred Forest.*
bottom left: Xyloolaena humbertii Cavaco; *middle: A local boy holds a seedling at the nursery near Ankarabolava-Agnakatrika; bottom right:* Breonadia salicina *(Rubiaceae) is a member of Madagascar's third-largest endemic family. This species is widespread in low to mid-elevation humid forest in the eastern part of the island.*

Opposite title page: *"Bonnet du Pape" or "La Brioche" in south-central Madagascar, near the Village of Zazafotsy*

Table of Contents page: *Tapia woodland at Ibity Massif*

Inside back cover: *A patchwork view of forest fragments and rice fields*

Back Cover: *top left: The nurseryman (Armand) at the Ankafobe forest transports seedlings of native plants from the nursery to the forest where they will be used for forest restoration; middle:* Phylloxylon xiphoclada—*a rare plant of the Malagasy highlands that is currently the focus of species-focused conservation action; right: Tree planting in the the yard of the Ankazobe High School*
bottom left: Flower of Kosteletzkya reflexiflora *(Malvaceae) in Andohahela national park; middle: Fruit of* Treculia madagascariensis *(in the fig family) at Makirovana-Tsihomanaomby; bottom right: Lower of* Xerophyta schatzii *(Velloziaceae), a new species restricted to southeastern Madagascar.*

FSC
www.fsc.org

RECYCLED
Paper made from recycled material
FSC® C020311

TABLE OF CONTENTS

FOREWORD

Dr. Peter Wyse Jackson

I am delighted to have the opportunity to write a foreword for this book that celebrates 25 years of achievements by the Missouri Botanical Garden in Madagascar. Although far from St. Louis, Madagascar has become a place where the Garden's global quest to document, safeguard, and raise awareness of the importance of plant diversity has become one of the most significant contributions that we make to the world. I am very pleased, therefore, that this book highlights so well why Madagascar is one of the Garden's top priorities in research, education, conservation, and in supporting the elusive goal of achieving sustainable development throughout the world.

The Garden has been fortunate to be able to collaborate as an equal partner with so many organizations and institutions, both in Madagascar and throughout the world. Such cooperation is an essential ingredient for this work, and I am pleased to thank and acknowledge all of them as well as our staff, colleagues, and supporters who have worked with us in Madagascar, many of them for more than two decades.

I first came to know of Madagascar when I was a child and watched the pioneering "Zoo Quest" television programs of David Attenborough. From that moment on, I was hooked with a fascination about the island and its spectacular diversity. While Madagascar is perhaps most popularly known for its fauna, such as lemurs—sifakas, indri, aye-aye—or the extinct elephant bird, the *Aepyornis*, its flora is equally remarkable, important, and diverse. During earlier parts of my career, through my work on the conservation of island plants, I would often talk about the amazing diversity of plants in Madagascar—10,000 vascular plants was the total I usually quoted. Now we know that number to be a very significant underestimate—14,000 is closer to the correct total, many of which have been discovered, named, and documented by the Garden's staff. Truly Madagascar is a global "hot spot" for biodiversity and one that is of paramount importance for the world.

In the earliest years of the Garden's involvement in Madagascar, exploration and discovery of plants was afforded the highest priority. Today that work continues undiminished, but it is made all the more urgent by the rapid loss of natural habitats, cleared through unsustainable agricultural practices, such as slash-and-burn cultivation, and as a result of many other human-derived causes. It is not surprising that the Garden needed to turn new attention to efforts to conserve plants and their habitats and support others in Madagascar, from government to local communities, to do so too. This book provides a wonderful synopsis and glimpse of some of this work. My congratulations and thanks are due to all those

FOREWORD *(continued)*

who have contributed to this book as well—providing information, data, stories, reminiscences, photographs, and so on. It is clear to me, too, that if the Garden had not taken up this challenge, much biodiversity in Madagascar would have been lost forever, with many species extinct or close to extinction. We also recognize that this vital work must continue now and into the future.

Madagascar may be far from St. Louis, but for the Garden it is a "home away from home." We look forward to many more years playing our part in helping to safeguard and cherish one of the greatest biodiversity treasures of the planet. Our pledge is to continue to work for sustainability with the people of Madagascar, to continue to support their efforts, and to help ensure that Madagascar achieves a secure future where people can live prosperous lives and co-exist in harmony with nature and the wild plants and animals with which they share this unique land.

Peter Wyse Jackson
President, Missouri Botanical Garden

Left: Clerodendrum petunioides Baker, Lamiaceae, *center: Baobab tree near Analavelona; Opposite page: Golden-crowned sifaka (*Propithecus tattersalli*), a species of lemur endemic to the Loky-Manambato region*

Top left: Sunset along National Highway 4; middle left: Bottle trees (Adansonia za) and rice fields with transitional vegetation (spiny and dry forests Andohahela corridor/Fort Dauphin); bottom left: Makay Massif in the southwest of Madagascar; top middle: Chameleon near Amboasary Atsimo; bottom middle: Goldrush at Ibity Massif—a conservation challenge; top right: Fruits of Cnestis at Makirovana-Tsihomanaomby

INTRODUCTION

Nearly any book or article one can find about Madagascar's natural history starts with some version of the following: Madagascar is a unique and incredible island off the coast of Africa, and is one of the foremost biodiversity "hot spots" on Earth. It may be that there are not many ways to express how fascinating the country is to any naturalist. It contains more plant diversity than Europe, and even though some other tropical regions may have more diversity, Madagascar tops them all in the number of endemic species on the island (90 percent). This is one of the reasons it is such an important research destination.

Although a map of the area would suggest otherwise, the island first broke from the African coastline when the supercontinent Gondwana began to split apart (between 184 and 135 million years ago (mya)), and then separated from India much later (between 88 and 65 mya). Madagascar is separated from Africa by the Mozambique Strait, roughly 300 miles wide, and it is thought that many of the island's animal species arrived after the split, floating on vast rafts of vegetation, some of which could stretch for several

Top: Gastrorchis sp.; bottom: Mouse lemur (Microcebus spp.)

miles. This geological history accounts, in part, for the unusual mixture of African and Asian plants and animals, and its geographic isolation and remarkable diversity of climate and geology are the reasons Madagascar has so many endemic species.

Some of the emblematic animal groups on the island include chameleons, tortoises, frogs, bats, fossa, and of course lemurs, an older branch of primates well-loved in zoos and nature parks around the world. Lemurs are found only in Madagascar, and may be the one reason many people even know about this island nation. But beyond their status as Madagascar's cutest inhabitant, they are also important to the plant population because they are great pollinators and seed dispersers. And, like many of the plants in Madagascar, many species of lemurs are endangered, some critically.

In addition to its unique fauna, Madagascar also boasts an incredible array of endemic plant species. Its endemic plant families include Asteropeiaceae, Sarcolaenaceae, and Sphaerosepalaceae, and there are thousands of species found nowhere else in the world (we believe there are approximately 14,000 vascular plant species). Among the more well-known

are the baobab tree, traveler's tree, and hundreds of orchids, of which over 80 percent are believed to be endemic.

Because it is a continental island, Madagascar provides a natural laboratory for studying its unique plant groups, many of which have existed there in isolation for millions of years. As the fourth largest island in the world, it has a remarkable variety of environments and habitats, including forests, deserts, mountains, swamps, and plains, all of which contain a rich diversity of plants and animals. It is also contained physically, making observations of its biodiversity—and changes to that diversity—easier. Nevertheless, Madagascar also has a fragile environment, susceptible to extreme weather and natural disasters like cyclones, flooding, and wildfires that threaten its many species.

The biggest threat to its biodiversity is its human population. Curiously, Madagascar has been inhabited by humans for a very brief time (estimated at between 2,000 and 1,200 years). It is believed that the first inhabitants came from present-day Indonesia, and they brought with them some of the agricultural practices that have been devastating to the plants and animals of the island. Madagascar's famed megafauna (large animals) are now extinct because of hunting and habitat destruction. These creatures included giant fossa, giant lemurs, and the elephant bird, which was over nine feet tall. Shifting or slash-and-burn cultivation (called *tavy* in Malagasy), used mostly for rice and sometimes other grain production, consumes huge swaths of forest to this day and is one of the principal causes (directly or indirectly) of deforestation. Loss of species is happening at an alarmingly rapid rate.

The government and many citizens of Madagascar are aware of the threats to its biodiversity, but poverty and economic instability force many people into an unsustainable relationship with the island's natural resources and environment. Since humans arrived in Madagascar, it is estimated that around 90 percent of its native habitats have been destroyed by slash-and-burn cultivation, timber extraction, charcoal production, and wildfires. Without conservation, it is possible that all the forests will disappear by 2025. It is this grim prognosis that motivates the Garden's research and conservation staff in

Madagascar to catalog the existing plants and habitats, observe changes to them, and above all and in partnership with the local communities, conserve what is left of this fragile environment.

This book is the result of a collaboration of the Garden's staff in Madagascar, Paris, and St. Louis, and offers a view inside the lives and work of the many dedicated individuals who make up our Madagascar Research and Conservation Program. The Garden has had a sustained research presence in Madagascar since the 1970s, and established a permanent base in the 1980s. That program grew locally and organically, and now there are over 150 staff members associated with this program, of which almost all are Malagasy. The activities of the program can be described using the five objectives of the United Nations' Global Strategy for Plant Conservation (GSPC): discovery, conservation, sustainable use, education, and capacity building. Indeed, these objectives form the structure of this book. Each staff member takes to heart the spirit of the GSPC, and their dedication— and that of their local community collaborators—provides the true strength of the program. *Misaotra betsaka!*

Helana   Sarcolaena grandiflora

Kakazo maventy 20m eo ho eo Tsy fahita ofa-Isy elo
Antelezana. Tampolo sy ny manodidina. Miavaka amin'
voany vaventy, miendrika kopy. Karazany iray amin'
ireo fanakaviam-ben'ny zavamaniry tsy fahita n'aiza
n'aiza eran-tany: SARCOLAENACEAE. Fanorenan-
trano no fena ampasaana azy ka fena landindomin-
doza ary mely ho fany taranaka raha tsy aravan-
tsika. Ka andeha re hifanome fanana hiaro azy.

*Counterclockwise from top left:
Field botanists begin their trek
into the forest; Uroplatus reptile at
Analalava; School children gather
in a classroom for a presentation;
Endemic and endangered plant
Sarcolaena grandiflora planted
and labeled at Antetezana village;
Sudden rains can come at any time,
and field botanists and collectors
learn to be prepared; Tsitondroina
waterfall at Ambalabe Vohibe Forest.*

GLOBAL STRATEGY OBJECTIVES

Plant conservation is not a new concept. For example, conservation of Deo-rahati (sacred groves) in India has ancient roots from the Vedic period, roughly 1700 to 110 BCE (Nipunage, D. S., Kulkarni, D. K.; Asian Agri-History; Apr-Jun 2010, Vol. 14 Issue 2, p185). However, an organized global approach to the problem was not formalized until 2002. The result was the United Nations' adoption of the Global Strategy for Plant Conservation (GSPC), an initiative of the U.N. Convention on Biological Diversity. The GSPC has been adopted by almost all of the world's countries, and its implementation is supported by a multitude of botanical and conservation organizations, including the Missouri Botanical Garden.

Plants are an essential resource for the planet. Many plant species have recognized practical uses such as food, clothing, shelter, and fuel. Likewise, many wild plants have economic and cultural potential in the form of future crops and commodities, especially given the challenges of environmental and climate change. Plants also help maintain environmental balance and ecosystem stability, and they provide habitats for the world's animal life. Although a complete inventory of the plants of

Left: Drosera madagascariensis–*a carnivorous plant; right: Conserving the forest also means conserving the rivers that run through them*
*Opposite page: Villagers meeting under the shade of a tamarind (*Tamarindus indica*) tree in Anadabolava*

PLANTS IN OUR LIVES

1. Plants are the foundation of all life.
2. There are roughly 400,000 vascular plant species in the world.
3. Every year, the equivalent of 31 million football fields of rainforest are cut down.
4. Plants have many uses for humans: food, clothing, shelter, fuel, medicine, oxygen production, and more.
5. By understanding the effects of climate change on plants, we can learn how it affects humans and other species.

the world does not yet exist, it is estimated that there are around 400,000 vascular plant species.

Many plant species and their habitats are threatened by climate change, over-exploitation, invasive species, pollution, and clearing for agriculture and other development. If this trend is not stopped and reversed, we will lose opportunities to discover and develop plants to solve very serious economic, social, health, and industrial problems. The conservation of plant diversity is of immediate concern to the indigenous and local communities where many of the threatened species and habitats occur, and these communities must have a voice in addressing its loss.

If the GSPC were fully implemented, societies around the world would be able to:
1. continue to rely on plants for basic necessities
2. harness the potential of plants to address pressing global needs and share the benefits equitably
3. learn to mitigate and adapt to climate change
4. diminish the risk of plant extinctions
5. safeguard the genetic diversity of plants
6. recognize, preserve, and maintain the knowledge, innovations, and practices of indigenous and local human communities that depend on plant diversity
7. understand that plants support our lives and that everyone has a role to play in plant conservation

The GSPC consists of five objectives which can be summarized this way: 1. Understand and document plant diversity, 2. Conserve plant diversity, 3. Use plants in sustainable ways, 4. Educate others about the importance of plant diversity and conservation, and 5. Build capacity to conserve and sustain plant diversity through investment and engagement.

Objective 1: Plant diversity is well understood, documented, and recognized.
Objective 2: Plant diversity is urgently and effectively conserved.
Objective 3: Plant diversity is used in a sustainable and equitable manner.
Objective 4: Education and awareness about plant diversity, its role in sustainable livelihoods, and importance to all life on Earth is promoted.
Objective 5: The capacities and public engagement necessary to implement the Strategy have been developed.

Objective 1: *Plant diversity is well understood, documented, and recognized.*
This objective directs us to document the plant diversity of the world and the uses and distribution of plants in the wild, in protected areas, and in *ex situ* collections. We must also monitor the status and trends in global plant diversity, identify plant species, habitats, and ecosystems at risk, and maintain a "red list" of endangered plants. We must make the data accessible to everyone and promote research in plant science and other sciences that touch on conservation of plant species and habitats.

Objective 2: *Plant diversity is urgently and effectively conserved.*
We need to plan for long-term conservation, management, and restoration of plants and plant communities, including ecosystems and habitats, preferably *in situ*, but where necessary, *ex situ*, preferably in the country of origin.

Objective 3: *Plant diversity is used in a sustainable and equitable manner.*
It is imperative that we control and halt unsustainable uses of plants and their habitats. We must instead promote development of more sustainable practices, while also giving consideration to the impact on individual and community livelihoods. We must be fair and equitable in sharing of any benefits arising from discovery of plant uses (medicines, compounds, etc.).

Objective 4: *Education and awareness about plant diversity, its role in sustainable livelihoods, and importance to all life on Earth is promoted.*
We must make sure individuals and their communities are aware of the importance of plant diversity and the need for its conservation and sustainable use. This effort must include mobilizing popular and political support.

Objective 5: *The capacities and public engagement necessary to implement the Strategy have been developed.*
Enhancing resources (human, infrastructure, financial) for plant conservation is also essential, as is linking local stakeholders and other interested parties in support of plant conservation.

Because they are presented in a linear way, one might assume that the five objectives are pursued sequentially, from 1 to 2 to 3 and so on. In fact, all are pursued both sequentially and simultaneously.

Without plants, there is no life. The functioning of the planet—and our survival—depends upon plants. The research and conservation efforts of the Missouri Botanical Garden in Madagascar are a true reflection of the vision and objectives of the GSPC.

This page, left: Analavory Geyser at Ampefy; top right: Species of spiraling liana at Vohibe Forest; bottom right: Local expert interview in Vohibe Forest

Opposite page, top left: White-fronted brown lemur at Analalava Forest; middle left: Teaching new technique of planting rice to villagers around Vohibe Forest; bottom left: Apocynaceae Juss; top right: Degraded patch of Vohibe Forest with reforested area in background; bottom middle: Ambalabe village outside of Vohibe Forest; bottom right: Mantidactylus lugubris hiding out on a tree trunk.

Plant diversity is well understood, documented, and recognized.

The Missouri Botanical Garden's research efforts in Madagascar (and everywhere we do research) have always centered on documenting and understanding plant diversity and assessment of conservation status. Understanding and documenting plant diversity take many forms; some of them would have been very familiar to Henry Shaw.

The cornerstone of understanding plant diversity is discovery, which itself is based on a nineteenth-century model of exploration. Going into an unknown habitat in search of new plant life is the *raison d'être* for many plant taxonomists. In all countries where researchers from the Missouri Botanical Garden work, we are on the front end of discovery in these often poorly known areas. The Garden is exceptionally good at this around the world; indeed, it is what we are best known for. Because of our expertise and the amount of time dedicated to discovery in Madagascar, Garden researchers have increased the estimated number of known plant species in the country from about 8,000 (the estimate in 1959) to between 13,000 and 14,000 native species.

Above left: Inventory of inselberg plants in Faliarivo, Anjomanakona Commune, Manandriana District; middle: Tristellateia madagascariensis; right: Trekking to the Analavelona Forest

The process of systematic discovery starts with a map—both literal and figurative—of all known species. Species names, locations, distributions, and total numbers are recorded and examined to get an overall image of the existing known plant diversity. Based on these mappings, researchers then look for gaps in the data that suggest the presence of additional species that have not been recorded in a particular location, or other information that may point to unknown locations or those where more discovery is needed.

Two main approaches to plant discovery and collection determine the next steps. *Monographic* studies focus on a single taxon. Plant taxonomists who engage in monographic studies seek to discover and describe all known plants in a single family or genus. The geographic distribution and size of the group help delimit the study, but the objective is to know as much about the members of the group as possible: most particularly their organization into subgroups and the characters that define these subgroups, but also their ecology, biology, and geography. *Floristic studies* use geographic location as the basis for discovery, and the objective is to list and then describe all the plants within a defined location, from a single forest to an entire country.

Because of its unique evolutionary history, Madagascar is home to an incredible number of endemic plant species. It is widely estimated that as much as 83 percent of the island's vascular plants are endemic. Some of the better known plants in Madagascar include baobab trees (100 percent endemic), orchids (87 percent endemic), ebonies and rosewood (nearly 100 percent), both of which are endangered, and many species of palm (99 percent endemic). Garden researchers have led the discovery of many new and newly

Left: Patrice making a collection of a strangling fig; middle: Dillenia triquetra; *top right: The supremely delicate flowers of* Talinella; *bottom right: Piper inventory in the Mandraka forest in the eastern part of Madagascar*

Crossing a river by car on the road to Fort Dauphin

Research collection sites are sometimes remote and can take hours or days to get to. Conditions are rugged, and there is a lot of equipment to carry in and out. Finding potable water can also be difficult. Sometimes researchers will bring water in containers, and luckily for them, it is customary in Madagascar to make a rice tea in the same pot used to cook the rice, thereby reducing water use and at the same time boiling (and sterilizing) the drinking water. And as is often the case, even the best laid plans can go awry, for example, when the collecting is so fruitful that the research teams are reluctant to leave the field. Our field botanist Patrice tells about a time when he and his team were collecting specimens in the Makira Forest, one of the biggest blocks of rainforest left in Madagascar. They had sent the porters ahead to set up camp and get a fire ready to cook their meal. Evening began to fall, and the field crew was hungry. They didn't know how far it was to the camp, and the porters had taken all the cooking equipment and water; however, the field staff had some rice and sweet potatoes in their packs. Collecting nearby water and using some pieces of aluminum corrugate normally reserved for pressing plant specimens, they crafted individual cooking vessels to make a meal. Patrice's wife is a herpetologist, so he also knew that there were edible frogs in the area, and mixed with the sweet potatoes, those became the main ingredient in their *laoka* (accompaniment to rice, known as *vary*).

described species, including over a dozen species in the genus *Schizolaena*, and have significantly increased the estimated number of ebonies in the last decade.

In addition to research on individual plant groups, taxonomic discovery is also conducted in 11 conservation sites around the island. These 11 sites (see p. 52) belong to a larger group of 79 sites identified in 2005 as Priority Areas for Plant Conservation (PAPCs) in Madagascar. Although originally most of our research was conducted in the forested regions on the east coast, renowned for their high species diversity, we now also conduct research and conservation in other lesser-known habitats and environments.

Exploration and discovery must have tangible results, and in plant science, those are the collections and the newly described taxa that are based on those specimens. In Madagascar, Garden-trained field botanists scout locations for collecting or return to sites that are still being surveyed. The botanical diversity of Madagascar is such that even sites that have been visited many times can yield new species. Whether looking for new species or confirmation of known species, botanists in the field gather the same types of data. First they will collect specimens from the plant: the leaf, stem, flower, and fruit (if available; if not, they will make a note to return when

those parts are present). Those specimens are given codes and pressed in the field, and will be sent to the office or the herbarium for further study. To accompany the plant specimens, the field researcher makes notes about the collection location including its name, geographical coordinates, vegetation type, geology, and also features of the plant that will not be evident in the specimen such as its growth form, the color of its parts, presence of latex and gums, and its odors. The accumulation and synthesis of specimen identifications from a particular location tell us a lot about the history of that location and its biological and physical environment.

When specimens have been collected and pressed, they are sent to the main office in Antananarivo (hereafter referred to as Tana) or to the herbarium at the Parc Tsimbazaza (PBZT), where a sustained partnership between our staff in Madagascar and the staff at PBZT's herbarium has created a botanical resource of national importance. Once identified, those specimens are entered first into the Garden's plant database, TROPICOS, and later into the Madagascar Catalogue of Vascular Plants. Multiple specimens are always collected so that one can be left in Madagascar's national herbarium at PBZT, while others can be sent to international herbaria including our own in St. Louis.

As the plant specimen is cataloged, the description includes the botanical name; the local vernacular name; ethnobotanical uses, if any; its conservation status, if known; and known efforts to conserve it. These data, along with the specimen and field notes, become the foundation upon which taxonomists begin the process of description and analysis. The specimens are compared with other collected specimens as well as published descriptions of species to verify either that the specimen belongs to the described species or that it represents a different or even new species. The findings are then published and become part of the permanent record in TROPICOS and the World Flora Online.

Madagascar presents many challenges to field research. Sometimes the sites are difficult to access. They may be in locations that are dangerous because of bandits, the presence of potent diseases, the absence of medical assistance, and treacherous routes. Some are inaccessible because of the season, road conditions, flooding, or other barriers, or they may just be too costly to get to in terms of money or time. Some sites on the west coast of Madagascar are accessible only by sea, and in many cases that means waiting, sometimes days, for transport on the ocean in an open boat from one point to the next.

From left to right Collecting specimens in Faliarivo; Barleria dulcis; Didierea madagascariensis *in a landscape near St. Augustin south of Tulear; Hairy mushroom in Andohaela Forest; Children living in the region of Andohahela Corridor/Fort Dauphin*

MR. EXCELLENT

Plants, like all other organisms, are classified by genus and species, among other categories. One of the reasons for collecting vernacular names of plants is that most people don't know the botanical name of plants, and instead call them by a common name. For example, a rose is a rose, but is also a *Rosa glauca* or a *Rosa rubiginosa* or a *Rosa* spp., depending on exactly *which* rose. But we may know them as American beauty or tea roses. Vernacular names thus have cultural significance. One day when I was out in the field with our technical advisor, Chris, he asked about the vernacular name of the palm *Dypsis carlsmithii*. When the forest guides responded, Chris's eyes grew big, and he did a double take. He asked them to repeat what they said, and, with everyone (except me) chuckling, he told me the vernacular name translated as Mr. (*Ra*) big (*be*) hitter (*doana*). I further Americanized it as Mr. Excellent.

An exceptional specimen of Rabedoana (Dypsis carlsmithii)

Once our researchers are in the field, there are additional challenges. Convincing the local communities that we are there to study the plant life is sometimes met with skepticism. Local residents may assume the unfamiliar field workers are there to look for gold or gemstones, to extract timber, or that they will interfere with local activities. One way we have addressed that skepticism is to try to do something with immediate tangible benefits for the local villages, like bringing a medical doctor along who is able not only to look after the botanists but also treat the villagers' illnesses at no cost to them. Building goodwill in turn can result in local assistance with the fieldwork. One of our best field botanists started his career as a cook for one of our researchers. His talents at identifying and collecting plant specimens were quickly recognized, and soon he was being trained as an assistant field botanist and then as a field botanist in his own right.

It is also important to learn the local *fady* or taboos, especially those associated with the collection area. *Fady* vary from one place to another, making it necessary to learn the specific ones of each new location, and presenting oneself to the village elder upon arrival is a good way to find out about them. *Fady* often include things to avoid like foods, times and places to walk or bathe, or wearing particular articles of clothing or colors. Hunting certain animals and disturbing the environment around sacred places are common *fady* in many locations.

Although it adds time to the process, including the local communities in collection activities builds relationships that can be long-lasting and beneficial for everyone. It sets the conditions for engaging the local residents in conservation efforts and facilitates understanding of the conditions and potential obstacles to carrying out those efforts.

Description and classification are among the products of all the exploration, discovery, mapping, collecting, list-making, and cataloging. Many of the staff in Madagascar are trained as taxonomists and are engaged in description and analysis. Comparing the existing information (publications and online databases) with the specimens (collections) can and does lead to more species identification. As previously mentioned, we have increased the estimated number of described native species, and for example, because of the research we have conducted in Madagascar, we know that ebonies are more diversified than anyone had imagined. Knowing the extent of a species' occurrence and its area(s) of distribution provide a baseline to use for understanding its conservation status. Besides conservation status, taxonomic analysis can help us establish the distribution of species, their origin, and their migration patterns. This information directly shapes conservation, education, capacity building efforts, climate change research, and our collaborations and consultancy work. Without the research, the rest is either not possible or not as grounded.

WHY CONSERVE BIODIVERSITY?

On my last evening in Madagascar, I asked Pete what I had started to call the "heretical question": *Why* is it important to conserve biodiversity? His response, paraphrased here, was as eloquent as it was powerful. First, there is what might be considered the pragmatic reason. Most plants have a direct or indirect utility. We need them to provide food, shelter, fuel, clothing, and so on. If we don't do something to conserve what we have, sooner or later we won't have anything at all. That concept is not always easy to communicate, especially when there is apparent abundance. So it becomes important to project scarcity into the future and ask people to imagine a world in which their children will not be able to live the same kind of life they live now, or that their parents lived. Most people want what's best for their children, and pushing them to consider the consequences of lost biodiversity for their children and their children's children can be a powerful incentive to live more sustainably. Another reason is more philosophical. Organisms live as parts of larger systems, they have evolved in concert with each other, and in many cases have symbiotic relationships. Humans are uniquely able to disrupt those systems in catastrophic and lasting ways. Yet we are also organisms, and as such, also must live as part of these larger systems. Destroying even a small part of any of these systems has repercussions that we may not even foresee, so we should be careful not to exert our power to disrupt, but instead be mindful of our part in the larger systematic universe. Finally, discovery and exploration are the foundations—but also the products—of the scientific process. Discovering a useful compound or characteristic of a plant comes as a result of finding, collecting, analyzing, and describing the plant, not as the reason for doing it. Almost no paradigmatic scientific discovery came as the result of looking for an end use; rather these results can almost be considered accidental. It is the curiosity, the desire to know, the systematic application of that scientific process that leads us— eventually—to more discoveries, more cures, more solutions to the problems we face.

Plant diversity is urgently and effectively conserved.

Above left: Transport by oxcart near Amboronabo commune; right: The baobab is one of the emblematic trees of Madagascar.

Ideally research—describing and understanding plant diversity—contributes to our understanding of the needs and priorities of plant conservation, which in turn informs policy formulation, which leads to conservation planning, which results in plant conservation. It is rarely that simple. Or linear. The research work that we do in Madagascar does, in fact, inform our conservation efforts, but the reverse is also true, and conservation efforts can expose us to previously unknown plants and habitats. Also, deforestation in Madagascar happens at such a rapid rate that it is often better to secure an area for conservation first and then conduct more robust research.

The Garden's research program in Madagascar was, for the first part of its history, primarily engaged in field collection and taxonomic analysis. Despite the warnings and predictions about habitat destruction and species loss, the Garden did not actively pursue conservation program planning or implementation. In fact, no botanical gardens were involved in organized conservation efforts, largely because that was the purview of conservation groups and NGOs. An informal discussion at the 16th

International Botanical Congress in St. Louis in 1999 led to the Global Strategy for Plant Conservation (GSPC), which was formally adopted in 2002. But it was Malagasy staff who took the lead in insisting that conservation should be an integral part of our program there. Ever since that time, the Garden's research efforts in Madagascar have included conservation.

The second objective of the GSPC is conservation. In theory it's an obvious corollary to plant research; in reality it poses real challenges to plant scientists who are often not formally trained to take on the often multi-faceted and overwhelming job. Nevertheless, because of their expertise and understanding of plants and their habitats, they can immediately comprehend the urgency of the need and are among the most passionate conservationists. The Garden tackles plant conservation in two ways: at the plant level and at the habitat level.

Plant conservation can be more difficult to undertake than that of animals. Animals are usually thought of as cute and they are easier to anthropomorphize. Because all other forms of life depend on them, plants are the unsung bedrock upon which all other life rests. But without a face, they tend to become a backdrop to everything else, a habitat for animals. Because of this, it can be harder to convince people to conserve plants. It can also be difficult to secure funding for plant conservation. Getting the attention of policy makers and the media—and therefore the public—can be a challenge.

Crucial to the formation of policy is developing realistic strategies to conserve plants and habitats. It's not enough to just say we want to save plant A; we have to figure out *how* to do that. In Madagascar, we work closely with government agencies at all levels to influence policy decisions and directions. We also work

Schizoleana tampoketsana

PLANT RESEARCH INFORMS PLANT CONSERVATION

George Schatz and Pete Lowry are taxonomists on staff at the Garden who have worked in Madagascar for over 30 years. In 1998 they realized that an existing collection of plant specimens constituted a new species, which they named *Schizolaena tampoketsana*, described and published in 1999. Other Garden researchers working in Madagascar, Chris Birkinshaw, Landy Razafindrakoto, and Mamisoa Andrianjafy, later discovered that this species was limited to a few individuals in two small patches of adjacent forest fragments and, as such, was in acute danger of extinction through deforestation if conservation measures were not taken. That forest became one of our 11 conservation sites. The plant was also propagated, and its seedlings were used to reinforce the wild population. Some seedlings were also nurtured in botanical gardens and planted in prominent public locations in the town close to the wild plants where they serve to increase awareness of the plight of this rare tree. Research and discovery led directly to plant conservation.

Chris Birkinshaw along with Mamy Rakotondrazaka and Jean François from MFG inspect the health of a seedling of a rare species at Park Ivoloina.

OUR WORK WITH MFG AT PARK IVOLOINA

Park Ivoloina is located near the port city of Toamasina (Tamatave) and is managed by the Madagascar Fauna Group (MFG). The MFG is a consortium of international zoological and botanical gardens, including, for example, the Garden and also the Saint Louis Zoo, where the Save the Lemur campaign is headquartered. There are two main objectives at Ivoloina: to preserve endangered Malagasy plants and animals that cannot be adequately conserved in the wild and to heighten awareness among local people of their natural heritage and how this may be conserved. Over the past six years our staff has worked with those from MFG to collect and propagate seeds of 23 critically endangered plants and to nurture the resulting seedlings within the park as a safety-net collection for the threatened wild populations. We hope that one day, when appropriate secure habitats have been located, some of these species can be returned to the wild where they truly belong.

closely with other agencies and NGOs such as the Conservation International, Famamby, Asity, World Wildlife Fund (WWF), Wildlife Conservation Society (WCS), the Rainforest Alliance, and the Madagascar Fauna Group (MFG), among others. The Garden has played a role in policy formation through our Priority Areas for Plant Conservation (PAPC) project. In consultation with the Malagasy government, we helped identify 79 unprotected PAPCs. In this regard, the Garden operates like a local NGO in Madagascar. Of the 79 PAPCs, the Garden supports conservation programs at 11 of them (see p. 52). We also collaborate with mining companies, national parks, and conservation NGOs to inventory and assess botanical diversity in their areas, and help them set up conservation areas. Without exception our work with the mining companies reveals the presence on the mine sites of threatened and/or localized endemic species that must be protected.

At the 11 conservation sites, we approach conservation by directly engaging the communities that surround these sites. As previously noted, we have hired local villagers to help with research activities. In the past, this was done on a more or less ad hoc basis—filling immediate needs as they arose. In the context of conservation, however, the efforts are much more structured and concentrated. Many of these 11 sites are not easily accessible from Tana, but more importantly we believe it is essential that conservation efforts be localized, so we empower a local staff to manage conservation at each of these sites. The local staff consists of a site-based conservation facilitator, forest guides, forest guards or patrols, and often one or more nurserymen. Different sites may include additional staff as needed (for example, the site at Analalava includes cooks when there are guests and school groups). These

Left: Cooks at Analalava sort and clean rice; right: Seedlings are raised in the nursery near Analalava for transplantation around the forest.

local staffs work closely with the community organizations, based in the villages or *fokontany* that surround the protected areas.

The site-based conservation facilitators are trained in plant science and/or conservation management at the university, and ideally they come from one of the villages that surround the protected site so that they already have roots in the community and are familiar with local customs and *fady*. The principal role of the facilitators is to advise and guide the community about conserving the site. They are also responsible for hiring and training the forest guides, while the local management committee hires the guards (and other staff). The facilitator is also the administrative officer for the site. They are in contact with each other and with the staff in Tana, but they mostly operate autonomously.

This last point is made even more important because of our reliance on another Malagasy cultural tradition, the *dina*. A *dina* is a social contract agreed on and enforced at the community level. Unlike a *fady*, which applies to groups of people in a more individual way, a *dina* is a set of rules and agreements among the entire community, with sanctions for breaking them. *Dina* can be about anything, but in this case, they are about conserving the forest by controlling activities that are potentially destructive to it. That can include not cutting down trees (or certain trees), not exploiting a resource except during a defined period, not hunting special animals, using fire as an agricultural tool only under certain defined conditions, not diverting water, and so on. Because ideally the site facilitator and staff all come from villages surrounding the protected area, they have a better sense of the kinds of prohibitions

Left: Evidence of successful reforestation at Analalava—five years ago this area was just a thick blanket of fern; right: The fish farm near Analalava, stocked with tilapia and royal carp.

FACILITATING CONSERVATION

The strength of our conservation program in Madagascar is that we work with local communities to foster a desire to conserve their natural resources and create or support conditions under which conservation is possible. In other words, we facilitate conservation, but we are not conservation managers. Ideally we empower the local communities to take action, and these efforts include some of the following:

1. Patrol the forest to control the use of natural resources.
2. Plant seedlings of native trees to restore degraded forests (160,000 in the last five years).
3. Monitor deforestation, reforestation, and other signs of change.
4. Build fire breaks around the protected areas and train local villagers to build and maintain them (43 km).
5. Establish plantations of fast-growing alien trees to provide local people with alternatives sources of timber and fuel.
6. Introduce new crops and new cultivation techniques that replace slash-and-burn agriculture (*tavy*).
7. Provide training and equipment for local residents to pursue economic alternatives to those that damage the environment, such as:
    a. Fish farming
    b. Vegetable gardening
    c. Houseplant cultivation
    d. Fruit tree cultivation
    e. Sewing, basketweaving, and other handicrafts

These are among the activities undertaken in the communities that surround the 11 conservation sites, sometimes in conjunction with other agencies, and sometimes just with our village staff. Some of these efforts prove more successful than others, but we never stop encouraging ideas to conserve the precious resources of Madagascar. Our most important activity, however, is supporting local communities in the sustainable use of their natural resources.

that can and will be followed and the kinds of sanctions that might result from breaking the *dina*. Together with conservation staff from the office in Tana, the local conservation facilitators, the local management committee, and local people create the rules that lead to forest conservation in each of these protected sites.

Our conservation program in Madagascar is based in part on scientific discovery and understanding of plants, but it recognizes the human effects on the environment and relies on local control and input to make conservation a sustainable community effort. Without the knowledge of what conservation priorities there are (or ought to be), efforts to halt and reverse the loss of plants and habitats would be unsystematic and potentially ineffective. However, without the committed efforts and partnerships with local villagers, other NGOs, and governmental authorities, we alone would be powerless to act.

Fire is one of the most destructive causes of deforestation in Madagascar, and one of the ways communities can prevent or minimize the damage is to construct firebreaks. With the help of the Garden, communities around 5 of the 11 conservation sites have built and maintain nearly 43 km (30 miles) of fire break. This photo shows fire break maintenance by villagers around Analavelona Forest.

Plant diversity is used in a sustainable and equitable manner.

Above: Diospyros sp.; middle: Child at Ambalabe; right: Rice fields along National Highway 2 between Antananarivo and Toamasina

Since human colonization of Madagascar, more than 90 percent of the county's natural ecosystems, including most of its forests, have been destroyed. Malagasy cut forests to clear land for rice cultivation, to make charcoal for fuel, and to use the timber domestically and commercially. Other causes of deforestation include the human-made, like mining and wildfires, and the natural, like cyclones. And in addition to destruction of habitats and the environment, loss of primary forest opens the door to aggressive, invasive alien species. Yet the Malagasy are some of the poorest people in the world, and they rely on the exploitation of natural resources for their livelihoods and indeed, in some cases, for their survival. The juxtaposition of these facts is at the heart of the problem of sustainable and equitable use of plants in Madagascar. Using plants sustainably—that is, in ways that do not risk the survival of the resource itself—can immediately affect the livelihood of vulnerable and disadvantaged people. Not doing so not only affects global and local biodiversity in terms of lost habitats and species, but will also affect human livelihoods in the future when the exploited resource is exhausted.

In the past, most conservation efforts were imposed from outside by foreign agencies or the national government and, for the most part, excluded local participants. Many of these efforts were unsuccessful because they were not considered legitimate by the local communities in which they were enacted and because they failed to take into account their real human needs. In the mid-1990s conservationists and land management agencies, as well as the central government, began relying on a Malagasy custom called the *dina*. As described previously, a *dina* is a local, community-wide agreement about certain social norms and prohibitions. Because they often involve land and resource use, *dina* came to be seen as an obvious tool for conservation efforts. However, a common mistake made by some conservationists, as well as special interest groups, was to try to fashion self-serving *dina* and then get the communities to adopt and enforce them (Andriamalala and Gardner, 2010.) Our staff in Madagascar understands that the very heart of a genuine *dina* is that it is democratically constructed. It must be agreed to by all because enforcement relies on potentially violating another

Malagasy concept, the *fihavanana*, or the sense of social cohesion of a community, and people may be reluctant to turn someone in for violating the *dina* for fear of disrupting that cohesion.

The experience we have gained over the last decade has convinced us that the only viable strategy for the conservation and sustainable use of resources in Madagascar is to truly engage the local communities and enlist them as allies in conservation efforts. There must be a mutually beneficial relationship based on mutually beneficial outcomes for both parties. Identifying such outcomes can be difficult to do because the majority of the population in Madagascar is rural and live by subsistence agriculture, and they rely heavily on the forests, marshes, and other vegetation types for their livelihoods. When population density is low this relationship is viable, but with an annual population growth in the country of three percent, burgeoning populations are exhausting natural goods and diminishing ecosystem services. We must help local communities improve their social and economic conditions in ways that do not destroy the habitats they rely on.

Left: Lantana camara, *a very invasive species; right: Mining sand for construction in the Ivoloina River outside of Toamasina.*

How Can We Help?

Some of the ways we help local communities use resources in ways that are sustainable include the following:

1. Identify alternative methods for planting staple crops such as rice and manioc to reduce or eliminate the need to cut and burn the forest.

2. Identify and provide appropriate new varieties of local food crops to increase yields and improve reliability of harvests.

3. Help local communities identify and secure funding (grants and loans) to finance community projects like building wells, schools, and nurseries, improving infrastructure, and funding site facilitators and other site employees.

4. Identify and enable villagers to adopt alternative sources of income that do not degrade the environment.

5. Educate people about the consequences of non-sustainable exploitation of natural resources so they can make informed decisions about their futures and the future of their children.

6. Help local people develop and implement realistic and acceptable plans for the sustainable use of their natural resources.

7. Reinforce and restore populations of over-exploited species.

8. Help restore the natural capital of degraded land so that it can support biodiversity and useful economic activity once again.

Opposite page top left: Armand with nurseryman and his family; middle left: Women's sewing collective near Vohibe; bottom left: Planting rice in Analavelona

Clockwise from left: Ring-tailed lemur (Lemur catta) at Anja Park; Corn dries outside a storage hut at Anadabolava; Woman at Ambalabe; Spondias dulcis "hog plum"; Digging holes to plant trees from tree nursery in Andohahela Corridor/Fort Dauphin

Left: Gervais, our guide at Analalava, demonstrates how to get water from the traveler's tree (Ravenala madagascariensis), so named because parched travelers can use this method to quench their thirst; right: Local builders are constructing an observation tower to attract tourists to see flying foxes (fruit bats or Pteropus) which roost in the trees across a small ravine from this site. It took us over an hour to hike to the site from the compound in Analalava, and these workers make that trip several times a day (with construction materials!).

ECOTOURISM

Analalava is one of the 11 conservation sites and is located on the eastern coast of Madagascar. Its proximity to the main port city, Toamasina (Tamatave), and a popular resort town, Foulpointe, makes it a potential candidate for development as an ecotourism site. Although the forest there is small (200 ha), it has five species of lemur (including an adorable newly described species of mouse lemur), a large roost of huge fruit bats (aka flying foxes), and a host of endemic plants, including exceptionally rare and attractive species of palms and orchids. Local builders are currently employed building an observation tower on a hillside that overlooks the bat roost. There are rustic but comfortable sleeping quarters, a dining area, and a classroom space that during my visit was being used by a group of 30 school children on an overnight outing. Currently only modest numbers of tourists visit this site, but nevertheless they provide employment for five guides and ten part-time cooks, all recruited from local villages. In addition, we plan that 50 percent of the income generated from visitors will be invested in local development projects. The tourism infrastructure is also used to support a thriving program of environmental education for school children.

The site is not very far from the Ivoloina Zoo, which is itself a tourist attraction. The smell of cloves from the local processing plants punctuates the air, and in late November through December, baskets of fresh lychees are stacked by the sides of the roads waiting to be taken to the port, delighting the eyes (and noses) of visiting tourists.

But the site is not without its problems. The road to Analalava from Foulpointe can be difficult to navigate, especially after a hard rain when the dips in the road resemble small ponds. Also, ecotourism is dependent on tourists and subject to the ebb and flow of that industry. Money for infrastructural improvements, such as road reconstruction, is limited.

The site facilitator and staff, both local and in Tana, are cautiously moving ahead, attracting more school groups and visitors to the site, knowing that they must remain flexible and able to adapt to changing conditions.

One of the ways we engage local communities, win their trust, and understand the issues that shape their lives is to make a long-term commitment to the area. We don't come in for two or three years and then leave when the funding runs out. Conservation is a central component to what we do in Madagascar, and therefore it is considered indispensable and not merely an add-on. In fact, we believe our conservation efforts are strengthened by the composition of our staff, which is almost entirely Malagasy, living and working in the place where the need for conservation is always in plain sight and clearly invested in the conservation of their own natural heritage. Both conservation and research staff members are passionate advocates for biodiversity conservation, and they work tirelessly to engage local communities, carry out programs, and overcome obstacles that can often seem overwhelming.

Reaching mutually beneficial outcomes is not always easy, but it would not be possible without complete transparency. We always tell the local community what our interests are: that we study plants and believe in the importance of botanical diversity; that we see important floras in their area; that we would like their help to protect it; and we would like to know how we can help them so they can help us. Local villagers may be skeptical of our claims, but our

*Left: Chairs and umbrellas are available for rent at the seashore resort of Foulpointe; right: A white form of the Rosy Periwinkle (*Catharanthus roseus*) is planted in a replica of a canoe near the entrance of Park Ivoloina. This species is native to Madagascar, but is now grown in all corners of the world as an ornamental and medicinal plant.*

ongoing commitment to the ecosystems we aim to conserve and valorize for the local community gives us time to build trusting relationships with local communities.

Securing employment for local people is one good way to build trust with a community and increase acceptance for conservation goals. Hiring site facilitators, guides, and nurserymen not only provides an income to those individuals and their families, but it also serves as a bridge between our staff and the local community, and turns a local villager into a de facto Garden staff member. The facilitator and other local staff members help us learn about the local *fady*, areas of local concern, and what's important to the local villagers. It also helps confirm that desired outcomes for the local community are consistent with the Garden's conservation outcomes.

Full-time conservation facilitators are embedded in the local community adjacent to each of the 11 conservation sites. These young, well-trained, and highly motivated Malagasy are tasked with enabling the local community to conserve their natural heritage through its sustainable use. They are empowered to define work plans in collaboration with local stakeholders, manage the implementation of the defined activities, make decisions, oversee budgets, and hire and manage teams. The facilitators are encouraged to communicate electronically amongst themselves and with support staff based in Tana, to share successes, challenges, and best practices. Occasional workshops are organized to allow face-to-face discussion. The main management structure at the site is one or more management committees that are composed of elected local stakeholders and are responsible for the implementation of the *dina*. We provide members of these committees with a

per-diem allowance to compensate them for the time they must invest in committee work, and we also provide them with a team of trained forest rangers who are responsible for patrolling the forest to detect infractions.

Community associations are created at each of the Garden's conservation sites to provide a group of interested people a forum for developing new economic activities and the work required to implement them. The nature of these economic activities varies by site and is dictated by local circumstances (for example having a market to sell produce) and local interests. Currently these activities include handicraft production; fish farms; cultivation of vegetables, houseplants, cloves, and coffee; pig farming; chicken farming; bee keeping; and clothes making. In these projects we try to provide opportunities for all sectors of society, regardless of their social position. These associations provide opportunities not only for economic improvement but also for promoting social cohesion and sharing information about the project.

Members of these associations have proved to be powerful local advocates for conservation who can express the benefits of conserving the forests in ways that only a long-time local could. They talk about cooler temperatures and plentiful clean water. They mention the importance of the forest for medicines and sacred objects like the *fisukina*, used in funeral ceremonies in the villages near Analalava. They worry about the dangers of fire and over harvesting timber, and they take pride in educating fellow villagers about ways to avoid both. Of course this local advocacy is and should be the goal of conservation in Madagascar.

From Lose-Lose To Win-Win

by Chris Birkinshaw, Technical Advisor to the
Missouri Botanical Garden's Madagascar program

The 800-hectare Oronjia Forest, a dry deciduous forest close to Antsiranana, in northern Madagascar, deserves to be conserved for its unique beauty: stout and extravagantly flowered baobabs and pachypodiums, rugged limestone cliffs, charming crowned lemurs, and brilliant white tropical birds—all framed by the emerald Indian Ocean. However, this forest is not just beautiful; it is also classified as a Priority Area for Plant Conservation on account of its diverse flora that includes a number of species that are known only from this site. Sadly, over the last few decades this botanical paradise has been degraded and continues to be threatened. Two major threats are the felling of trees by impoverished local people for charcoal production and invasion by non-native tree species. Charcoal is used widely in Madagascar for cooking, and its production provides the livelihoods for many poor people. The destruction of this forest would be a catastrophe for both biodiversity and the local economy that relies on tourism. Recently two of the Garden's Malagasy staff members, Jimmy Razafitsalama and Christian Claude, have completed the implementation of a pilot project that aims to provide new and better livelihoods for the charcoal producers while simultaneously reducing the abundance of the alien trees. With the support of the Marisla Foundation, three experienced wood carvers from Ambositra, the famous artisanal town of central Madagascar, were selected, hired, taken to the Oronjia, and tasked with training 15 local charcoal producers to carve an array of animal models using the wood from of two of the alien tree species. The initial training focused on the production of relatively simple models that can be produced successfully even by new carvers but that are nevertheless judged to be attractive to passing tourists. After a few short months the trainees are already making and selling an array of delightfully naïve animal models, and in doing so have reduced the abundance of some of Oronjia's alien trees. That is what we call a win-win scenario!

Clockwise from top: Trainees practice their newly acquired skills in wood carving; Instructions for making a zebu candle holder from a branch; The final product and the promise of a better livelihood; A charcoal producer initiates a new model.

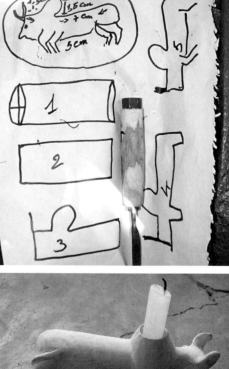

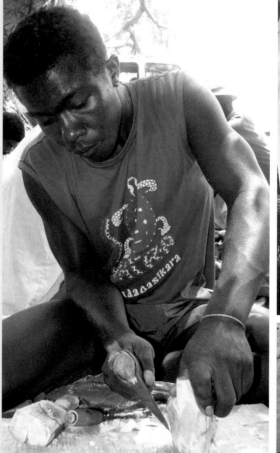

Education and awareness about plant diversity, its role in sustainable livelihoods, and importance to all life on Earth is promoted.

Above: left: Planting rice outside Vohibe; middle: "Twinning" Schizolaena laurina at Ambila Lemaitso; right: A schoolgirl recites a story depicted on the wall of her school in Firarazana near Ankafobe

In Madagascar conservation tends to focus on animals such as lemurs and tortoises. At the Missouri Botanical Garden, we are one of the few organizations that focuses on plants, although our work does overlap considerably with other organizations. Because, of course, Malagasy animals rely on plants for food and shelter, there is a natural alliance between plant conservation and animal conservation, such as our work with the Madagascar Fauna Group (see box on p. 28).

Our conservation partners often work in their own designated areas, sometime with our assistance, but our specific conservation efforts in Madagascar are currently based in the 11 conservation sites (see p. 52) that are scattered around the island. We selected these 11 from among the 79 identified Priority Areas for Plant Conservation (PAPCs) primarily for their exceptional importance for plant conservation and because they would not have been targeted for conservation action on the basis of other considerations, such as the presence of emblematic animals. All contain an unusually high abundance of plant species that are not known from any other protected area.

In Madagascar we don't just "do conservation;" we teach by example and use our own practices to educate and raise awareness. Our success in conservation is tied to our practice of enlisting community support and engagement by building awareness about the importance of conserving biodiversity. We work closely with local communities that surround the protected areas, helping them organize community associations tasked for various conservation work, and training and securing employment for local site facilitators, guides, patrols, field researchers, and so on. We treat all local staff as potential advocates for conservation and train them accordingly. In addition to completing their specified tasks, we want them to believe in the project's goal. In doing this, our conservation projects are not only focused on conservation today but also on nurturing the next generation of decision makers. Our local staff includes many younger people, some who have families and some who don't. They are not yet village elders, but training them now in the importance of conservation helps assure that when they attain this position, they will understand and promote sustainable use of plants and the need to conserve biodiversity.

In addition to training local workers, we are also helping the next generation have a different vision for their future, one that includes a conservation world view. At the university level, we train graduate students in various aspects of applied conservation with a particular current emphasis on ecosystem restoration. Our conservation sites become the field sites for teaching and research, and they are ideal locations to establish secure experiments on controlling alien invasive plants and propagating and growing native plants. Other graduate students are being trained in taxonomy, ecology, ethnobotany, and floristic studies. Over the years we have trained 3 doctoral and 66 master's students. We have also trained a generation of field botanists who work with our researchers to help with collecting and who also help train local field workers, and we train outside consultants who work with mining companies to establish conservation areas to preserve the plants that would otherwise be lost to mining activities. It would not be an exaggeration to say that if one were to meet an active field botanist in Madagascar, he or she would almost certainly have been trained by our program.

Left: The nursery at Ankafobe; right: Garden staff members hike up the hillside from the forest fragment in the valley at Ankafobe during a staff outing.

A group of school children on an overnight camp trip to Analalava Forest play games in the field outside the classroom building. In this photo, the children are taking turns entering the circle to perform a rhythmic and slightly silly "kestrel dance." This activity produced much laughter and hilarity, and allowed the children to burn up a little energy!

Kestrel Dancing in Analalava

We pulled up to the forest site in Analalava on a Friday in mid-morning and were greeted by the sight of a large group of young children. I had been told the evening before that there would be a school group on an overnight outing while we were visiting the site. The kids were playing a circle game when we arrived, but soon broke up into small groups of ten or so. Each group was paired with one of the forest guides, who on this day acted more like camp counselors than docents. The three groups went to different locations around the central field, where the guides led them in different games, all focused on raising awareness of their natural heritage. One group in particular was busy learning the names of five rare and locally endemic plant species found in the forest just a stone's throw away. The guide held up a specimen, and one by one each pupil was required to name them all. There was much teasing for incorrect identification and cheers for the correct answers. When all five had been correctly identified, the kids all shouted a joint congratulatory cheer. After this activity, the whole group reunited in the center of the field to play another game. This time the kids stood clapping in a large circle, and in groups of two or three, they entered the center of the ring, hands on hips, to perform a "kestrel dance." It was an adorable sight! The afternoon was spent hiking in the forest to see *in situ* the plants and animals that had been discussed that morning, followed by some quiet time, dinner, and then a night hike to see the nocturnal mouse lemur. The evening was capped off with a campfire complete with songs, stories, and an energetic game of "telephone." Millions of stars filled the night sky, and laughter filled the air. The day had been a wonderful experience that combined the fun of camp with important lessons about conserving their own forest.

Public education efforts include outreach programs, radio broadcasts, newsletters, and many school programs. We work with primary and secondary schools to incorporate environmental awareness-raising into the curriculum. Some of the conservation sites have overnight "camps" for primary school children led by the forest guides with help from the school's teachers. Through games and challenges, the children are taught to know and enjoy natural places and the plants and animals that live there. They typically take a hike in the forest, play games, and tell stories at night. This program is in its early stages, so there are many great ideas for expanding it, possibly into something that resembles the scouting program in the U.S.

Another community-level education program is one that the staff in Tana refers to as "twinning." This involves pairing a village with a particular species of plant—one that has a special status (it is endangered; it is endemic to the locality; etc.). The plant then becomes like a sort of mascot for the community. They plant seedlings of their adopted species in prominent public places with a sign informing passersby of the importance of the plant. The day when the seedlings are planted is a day of celebration during which a range of events (quizzes, theatre, dancing, speeches, and poetry reading) celebrating the village's special plant are held. One example of this is the twinning of the *Schizolaena tampoketsana* with the residents of Ankazobe, the nearest town to Ankafobe Forest where the last population of this endangered tree has its stronghold. This tree is now so appreciated by the people of this village that they have integrated drawings of its fruits into the town's shield.

Left: The forest guides at Analalava, Jeriste, Michel, Gervais, and Fahede; top right: A guide leads the children in a plant-identification exercise; bottom right: The children and guides enjoyed a campfire complete with songs, stories, and a game of "telephone."

Outreach is a necessary component of conversation, but it is not always easy. For example, many Malagasy rely heavily on forest clearing for their livelihood, and without viable alternative means of generating livelihoods, the short-term cost of conservation often overpowers its long-term benefits. Therefore investment in raising environmental awareness is largely wasted unless local stakeholders have access to alternative economic options from those that degrade the environment. When such options exist, then outreach is most effective when it has a broad focus and when the long-term economic implications are made explicit. What, besides plants, is worth conserving in the forest? The forests are home not only to endangered plants but also to threatened animal species, which can attract tourists and which may be considered protected by local *fady*. The forest usually contains traditional medicines as well as burial grounds, so the surrounding forest is considered sacred. Also, the water that people rely on for drinking and cooking, not to mention irrigation, washing and bathing, construction, and so on, comes from the forest. The villagers understand that, and it is not difficult to point out that a loss of forest means a loss of all of these things. The role of the Garden is to identify the most important and threatened habitats, educate about the importance of conservation, help identify alternatives, support local efforts to use resources sustainably, promote and facilitate local sustainable economic development, and serve as allies in the national struggle to stop the rapid deforestation on this beautiful island.

Top left: The school and pupils at Firarazana near Ankafobe; bottom left: The Garden staff members enjoy a picnic lunch in a gazebo just outside the forest at Ankafobe; right: With help from the head of Firarazana village, the author was invited to plant a Schizoleana tampoketsana seedling as part of a reforestation effort while the Garden staff watched (presumably to make sure it was done properly!).

SEEDLINGS AND SOCCER BALLS

The forest at Ankafobe is a mere 130 ha, surrounded by degraded grassland that is prone to wildfires. The forest was once much more extensive, as is testified by singular trees now standing alone far from the actual forest edge. Also the grassland in this area is secondary and very species-poor; there seems to be very little native grassland on the Malagasy Highlands. In an effort to stop further deforestation and reverse the trend, the Garden has teamed up with the local village of Firarazana to plant seedlings of a variety of native tree species—on the order of 5,000 per year. The site is on a major highway that links Tana with the west coast city of Mahajanga, and as such, is a potential stopping place for tour busses and other drivers making that half-day journey. Seedlings of *Schizolaena* and other trees dot the hillsides between the forest and the village, and the hope is that in a few years, the landscape will look quite different.

On my second-to-last day in Madagascar, the Garden staff had an outing to Ankafobe to plant some of those 5,000 seedlings as part of the forest restoration project, take the short hike through the forest, and enjoy a picnic lunch. After lunch we returned to the village where the school children were gathered outside their new school (built with the support of the Garden). There was a mural on the wall facing the school yard, designed and painted by staff member Roger

Andriamiarisoa. It depicted stories that had been collected from the village elders about local cultural and natural heritage. The children had been learning the stories illustrated by the pictures on the wall, and some had been enlisted to tell them to us. Using a long stick, different students stood up one by one to recount the stories while pointing to its illustration. Following their performance, the villagers headed over to a big field for the main attraction of the day: a football (soccer) match between the Missouri Botanical Garden team and the local team that was composed of the men who work with us during the year to make firebreaks and to grow and plant tree seedlings. The Garden team had what should have been a successful strategy: make the other team laugh so hard they couldn't play. Even when our goalie took a call on his phone in the middle of the game, the village team stayed focused and won the game 2-0, taking home a 20-kilo sack of rice and a new soccer ball as a prize.

Engaging local communities on many levels from children up to older adults is a key factor in successfully enlisting their help with conservation. The relationship must have mutual benefits for it to continue. And because conservation is an ongoing concern, the durability of those relationships is critical to our conservation work in Madagascar.

Left: Richard Randrianaivo talks to the nurseryman, Armand, at Ankafobe while the seedlings are being exposed to full sun prior to planting; right: The team photo just before the match between MBG and the local forest workers. The local team beat us 2-0.

The capacities and public engagement necessary to implement the Strategy have been developed.

Above left: Armand, Tefy, and field botanists at Analavelona; middle: Flowers of one of the many new species of ebony (Diospyros) being described from Madagascar; right: Nivo Rakoarivelo and Rondro Ramananjanahary working at FOFIFA, one of Madagascar's two national herbaria.

Capacity building has been a key component of international aid and development programs for several decades, and it has different applications depending on the development goal. In the case of conservation, the goal is to enable local stakeholders to develop sustainable ways to use resources, including preventing its unsustainable use, substitution and/or crop diversification, and changes in economic output that shift the burden away from dwindling resources. Capacity building therefore refers to both understanding the obstacles to a sustainable use of resources and enhancing the local populations' abilities to conceive and implement actions to reduce these obstacles. Deforestation is a serious concern for everyone in Madagascar, but the obstacles to stopping it are equally serious and can be difficult to address. A long-term and sustainable commitment to identifying and addressing the root causes of actions that lead to deforestation is needed. This is where capacity building enters the picture.

Its ties to large-scale development on a national level give the impression that capacity building happens only on a similarly large scale, but in fact it happens at all levels of organization. In Madagascar, the Garden's involvement in capacity building happens at three levels: the community level, which is associated with our 11 conservation sites; training the next generation of taxonomists, botanists, and conservationists and providing them with work experience opportunities after graduation; and supporting activities at Madagascar's two national herbaria. Perhaps the most important level at which we have built capacity is within our own local staff, who are trained professionals with a level of knowledge, commitment, and influence that enables them to make a real difference. This is directly the result of the Garden's internal capacity building.

At the community level we hire people to satisfy an array of functions required for effective conservation: managers, rangers, nurserymen, guides, and education officers. Once trained these people are exceptionally valuable because they serve as liaisons between the Garden and the village, understanding the interests of both parties while upholding the conservation mission of the Garden. Through training, continuing education, and a continual exposure to a conservationist worldview, these individuals are better able both to understand the profound implications of a loss of biodiversity and to work effectively with their communities to come up with realistic alternatives to practices that contribute to it. Often these local staff members hold positions of some importance within their villages and can be a powerful force for changing the perceptions of local people and their natural ecosystems.

To a greater or lesser degree our conservation projects are both managed and implemented by the local communities themselves, and the role of our permanent staff is to direct, facilitate, and monitor these actions. Nearly all of our permanent staff are Malagasy and therefore understand the realities of achieving conservation in Madagascar.

Left: Field collectors pack a canoe during a collection trip; middle: Establishment of tree nursery at Ankoba/ Fort Dauphin; top right: The author learns about ex situ *forest conservation at Park Ivoloina; bottom right: Haingo and Nivo after planting seedlings at Ankafobe*

Armand Randrianasolo was a student at the university in Tana in 1989, pursuing a master's degree in biochemistry. One day his professor told the class that some American botanists wanted to interview student volunteers to help with plant collection in the forest. Armand was selected and shortly after introduced to Jim Miller, then the director of the William L. Brown Center at the Missouri Botanical Garden. Without proper footwear or a backpack, Armand left for the field the following day, using Jim's spare pack and wearing a new pair of boots that the program had purchased for him. For 15 days Jim taught Armand plant identification, and then quizzed him on what he had learned. After this field trip, Armand met Pete Lowry in the office in Tana, and all agreed that Armand would make an excellent botany student. He became the first student trained by the Garden in Madagascar. He started out in the herbarium in Tana at Parc Tsimbazaza, then left to study in St. Louis where he earned his MA then his PhD. He went on to complete a post doc with Jim Miller in St. Louis, and then was hired by the Brown Center, where he now specializes in the flora of Madagascar. Armand now supervises his own graduate students in Madagascar, trains field botanists, and collaborates closely with the staff in Tana. This may be the first, but it is not the only, story of building capacity in Madagascar.

Although education is not the only tactic for capacity building, it remains an important one. Over the years in Madagascar we, in collaboration with various Malagasy universities, have trained more than 60 Malagasy master's degree students and have helped them obtain employment that makes good use of their acquired knowledge: in conservation, government, NGOs, and industry. The subjects for the master's studies commonly include taxonomy, ecology, restoration, and an array of themes related to applied conservation management. Many more young Malagasy have received technical training, and we are particularly proud of the work we have done over many decades to train scores of productive local plant collectors. Many of those we have trained are now acting as trainers themselves.

The Garden's professional staff in Madagascar has been built from the inside out. As mentioned previously, most of the staff is Malagasy, and those working at the our conservation sites are recruited from local areas, trained at the university, then they return to local sites making it more likely that they will stay there. In fact, in our conservation and capacity building we act as a sort of in-country NGO and are asked to assist with capacity building on a national level.

In our meetings with local community associations, one topic that arises frequently is development. Thanks to the involvement of the Garden, village elders who head up the community associations are committed to conservation and see the benefits of sustainably using the forest resources. However, they also answer to their villages, whose residents rely on the forest and whose traditional practices are not sustainable but often can't see any alternatives. Association leaders work with the local forest managers and staff in Tana to propose alternative sources of income as well as more sustainable ways to manage the forest resources. Some of these development ideas, proposed and pursued by the community associations, include ecotourism; fish farming; raising vegetables, houseplants, and other less land-intensive crops; and crafts such as carving, basket making, and sewing. The Garden can help find funding for start-up projects, can help villagers write grant applications, and can help identify markets and distributors for the crafts created in the villages, but the ideas and implementation come from the villagers. Our goal is to serve as a liaison, not the principal, in economic development; that way, the community associations become self-reliant as they navigate the process of raising money and starting new industries. Community associations established for purposes of conservation are thereby transformed to include an entrepreneurial facet as they pursue economic opportunities that replace practices that are damaging to the environment.

Our long history of commitment to the country makes us one of the important stakeholders in national conservation efforts. Our collaborations with other NGOs, companies, universities, and government agencies allow us to influence research and conservation in lots of areas. Our 11 conservation sites are jointly managed with local communities. And one of our true strengths is our approach to conservation and capacity building: we think of it more in terms of relationship building, with the ultimate goal being a benefit to the communities we work with. We invest years in building and maintaining collaborative establishing relationships with the communities surrounding the conservation sites, building trust and mutual respect over that time, and the reward is evident in the pride of the villagers cultivating their own vegetable gardens, earning market value for handicrafts, and singing songs they have written specially to honor our staff at the various biodiversity festivals we jointly hold.

Left: The school assembles to watch the planting of a seedling of Schizolaena tampoketsana *in their yard; right: Malagasy landscape at the Anja Park.*

This page, top left: Pupils practiced the lesson learned about pressing plants in Vohibe;
top right: Ascension of inselberg in Anja Park;
right: Garden restoration expert, Cyprien Miandrimanana (center), with students from the University of Toamasina (Tamatave), gathering for a break from their research at Analalava
Opposite page, top: Nadiah Manjato and Ornella Randriambololomamonjy working at the FOFIFA herbarium, bottom: Nadiah Manzato and Ando Ramahefaharivelo in the field

How Do We Build Capacity in Madagascar?

- We work with universities in Madagascar to support and train students in ecology, taxonomy, and applied conservation management.

  – To date, 88 students have graduated with our support, 66 with master's degrees.

  – Currently 85 percent of the graduates are engaged in careers that make full use of their training.

- We have long supported Madagascar's two national herbaria with improving infrastructure (critical roof repairs to the existing herbarium building in 2003 and construction of extension in 2005 to double its capacity), equipment, and supplies.

- We provide professional opportunities to the staff members of the herbarium.

- We provide horticultural equipment (labeling machine, mowers, chain saws) and display elements (shade house and rock gardens) for the Parc Tsimbazaza Botanical Garden.

- We provide short-duration on-the-job training for our staff and the staff of our partner organizations in diverse subjects including, for example, field botany, estimating risk of extinction, and leadership.

- We empower local people adjacent to our conservation sites to implement conservation action by providing training and then coaching in diverse subjects including, for example, nursery techniques, good governance, leadership, English language, conservation education for local teachers, and monitoring indicators.

- We contribute to primary and secondary curriculum development.

MISSOURI BOTANICAL GARDEN'S CONSERVATION SITES

This map indicates the locations of all 11 of the Garden's conservation sites. Most of them are in the eastern part of the country, but we also have established protected areas in the south, north, and Malagasy Highlands. The photos to the right were all taken at each of the sites.

Oronjia

Makirovana-Tsihomanaomby

Point à Larrée

Analalava

Ankafobe

Vohibe

Ibity Massif

Analavelona

Agnalazaha

Ankarabolava-Agnakatrika

Anadabolava-Betsimalaho

ANKAFOBE FOREST

IBITY MASSIF

ORONJIA

MAKIROVANA-ANJAMBALAVA COMPLEX

AGNALAZAHA FOREST

ANALAVELONA SACRED FOREST

ANALALAVA FOREST

ANKARABOLAVA-AGNAKATRIKA

POINTE À LARRÉE

VOHIBE FOREST

ANADABOLAVA-BETSIMALAHO COMPLEX

OVERVIEW OF THE 11 CONSERVATION SITES

The Missouri Botanical Garden's Madagascar program has identified and developed community-based projects at 11 priority areas for plant conservation throughout the country. To enable conservation through the sustainable use of natural resources by local people, we place well-trained and motivated Malagasy conservation facilitators within the communities next to the site. Their responsibilities include working with local stakeholders to negotiate a mutually acceptable boundary for the conservation site, and developing and implementing local traditional rules (*dina*) that control the management and use of the area's natural resources. We raise awareness among local stakeholders regarding environmental issues so that they can make informed decisions. When access to natural resources is restricted, we work with the local communities to develop alternatives. As poverty is the root cause for most non-sustainable exploitation of natural resources in Madagascar, we also partner with development organizations and commercial interests to promote appropriate activities in the communities surrounding the conservation sites.

Prototypes of hand woven baskets for sale to the Blessing Basket Project—opening a new market for the people living around the Agnalazaha Forest

Many of the 11 conservation sites are located in the eastern regions of the island, where most of the remaining forests are located, but sites in other parts of the country have been developed as well. Each of the 11 sites represents a unique case for conservation, and in addition to forests, they include woodland, natural shrubland, rock vegetations, marshes, and swamps.

Agnalazaha Forest, on the southeast coast, is a 2,250-hectare fragment of rare littoral forest, with some marshes and lakes. It contains several locally endemic plants as well as the critically endangered gray-headed lemur (*Eulemur cinereiceps*). It is heavily used by local people for construction timber and for medicines, foods, fuel wood, and materials for handicrafts.

Anadabolava-Betsimalaho Thicket in the southern portion of Madagascar is 18,100 hectares of subarid deciduous thicket on basalt in a habitat known as spiny forest. It contains many drought-adapted plant species, birds, and lemurs, including the ring-tailed lemur (*Lemur catta*), the Verreaux's sifaka (*Propithecus verreauxi*), and mouse lemurs (*Microcebus* spp.). The area is the source of the Mandare River, which is the only source of water for hundreds of thousands of vulnerable people.

Analalava Forest is a very small (204 ha) low-elevation evergreen forest fragment along the eastern coast. It is botanically diverse, with many threatened plants, a remarkable 26 species of palm, and 10 species of plants that are now known from nowhere else. It is home to five species of lemur and has great potential as a destination for ecotourism.

Analavelona Sacred Forest, in the southwest, is a 14,400-hectare portion of rare subhumid/humid forest on basalt located in in a subarid landscape. It has very high local endemism and is the habitat of the red-fronted brown lemur (*Eulemur rufus*), the red-tailed sportive lemur (*Lepilemur ruficaudatus*), and the Madagascan flying fox (*Pteropus rufus*). It is also the source of the Fiherena River that feeds extensive areas of rice fields.

Ankafobe Forest is a tiny (130 ha), rare mid-elevation humid forest fragment in the Malagasy Highlands in central Madagascar. It contains much of the world's remaining population of the critically endangered tree *Schizolaena tampoketsana*.

Ankarabolava-Agnakatrika Forest, near the southeast coast, comprises 2,400 hectares of low-elevation humid forest on basalt and is the last area remaining forest in the district. It has a highly diverse flora with many locally endemic species, and six species of lemur including the gray-headed lemur (*Eulemur cinereiceps*) and the collared brown lemur (*Eulemur collaris*).

Ibity Massif, in the central highlands, contains 5,630 hectares of sclerophyllous woodland, shrubland, herb-rich grassland, and gallery forest, and is especially important because of a score of plants that are only known from this site and because it includes vegetation types that are not well represented in existing protected areas. The site has good potential for tourism.

Makirovana-Tsihomanaomby Forest near the northeast coast is a 5,200-hectare low- and mid-elevation humid evergreen forest on basalt and laterite. The diverse flora includes local endemics and is the habitat of the endangered Sanford's brown lemur (*Eulemur sanfordi*) and the crowned lemur (*Eulemur coronatus*). The forest exists in a transitional bioclimate from subhumid to dry and contains the sources of several rivers and streams that irrigate extensive rice fields on the adjacent plains.

Oronjia is in the far north of Madagascar. It is a 1,500-hectare dry deciduous forest on sand and calcareous rock with vegetation types that are currently under-represented in the protected area network. It contains threatened and locally endemic plants and is heavily exploited for charcoal production. Because of its location, adjacent to the beautiful Diego Bay, it has great potential for tourism.

Pointe à Larrée, on the eastern coast near the famous tourist island of Île Ste Marie, is 5,330 hectares of low-elevation forest on laterite, littoral forests, marshes, swamps, and lakes. This mosaic of vegetation types contains considerable botanical diversity with several threatened and locally endemic plants. There are many species of birds and six species of lemur. This site has huge potential to generate income for local people from tourism.

Vohibe Forest is a 3,117-hectare low-elevation humid forest in east-central Madagascar. This forest is valuable because little low-elevation forest now survives in Madagascar, and this forest is an almost pristine example of this rare vegetation type. To date 350 species of plant have been recorded from the site, as well as 11 species of lemur including the eerie-sounding indri (*Indri indri*). The site boasts one of Madagascar's biggest waterfalls and has some potential for adventure tourism.

The grass just outside the meeting rooms/classrooms at Analalava has been sculpted into the shape of the MBG logo.

*Clockwise from top left:
Craterispermum Benth; Researchers
stuck in the mud near Ankorefo/
Antsiranana; A view through the
palms in Vohibe Forest; Tsitondroina
waterfall in Vohibe; Local nurseryman
with Mantalania longipedunculata
at Park Ivoloina (as part of the ex
situ conservation project); Richard
Razakamala cutting samples from a
tree; Psychotria retiphlebia; A villager
from Ambinanitelo Sakaraha in
southwest Madagascar; Tefy Harison
Andriamihajarivo asks, "should I eat
these fruits or press them?"*

Agnalazaha Forest

This forest is one of the largest fragments of littoral forest (defined as low-elevation humid evergreen forest on sand) remaining in Madagascar. The littoral forests of eastern Madagascar have rapidly shrunk since the island was colonized by humans between 1,200 and 2,000 years ago, and the few remaining patches are in danger of disappearing before we have a clear idea of their species composition. It is widely believed that they once formed a continuous band along much of the eastern coast of the island; however, the littoral forest now exists as isolated fragments which are under increasing pressure from local agricultural practices, the impact of cyclones, and fire.

The Agnalazaha Forest is located in southeastern Madagascar, about 50 km south of Farafangana, within Mahabo-Mananivo Commune (see map on p. 52). Our inventories of the forest and adjacent marshes, rivers, and lakes indicate a rich flora and fauna, with several species that are locally endemic and endangered, such as the rare and threatened gray-headed lemur (*Eulemur cinereiceps*).

Conservation efforts in this forest include ecological restoration through eradication of invasive plants and reforestation using locally propagated seedlings of native trees. Wildfires are frequent in the wind-blown coastal plains, and the forest is protected from burning by means of firebreaks. Staff and students conduct experiments to compare survival and growth rates of seedlings of different tree species used in restoration at this site.

Among the development projects undertaken in the communities adjacent to the forest are vegetable gardening, higher-yielding rice cultivation, and maintenance of irrigation systems. We collaborate with the Blessing Basket Project to enable local weavers to produce attractive and high-quality baskets for the sale in the United States.

Agnalazaha local partners:
Mahabo Mananivo Commune, Soazagnahary

Left: Agnalazaha is a littoral forest with a white sand floor; top: Community-based conservation effort is a Garden-supported nursery in Mahabo which provides saplings of fast-growing trees as an alternative timber source to the native forest trees; bottom: Family portrait in Agnalazaha.

Anadabolava-Betsimalaho Thicket

The Anadabolava-Betsimalaho Thicket in southern Madagascar is important both for its rich biodiversity that includes several locally endemic plant and animal species and for its role in protecting the watershed of the Mandrare River, the principal source of water for 350,000 people. However, like many of our conservation sites, this site is threatened by shifting cultivation, locally called *tavy* or *hatsaka*, that currently constitutes an important means of livelihood for the local population. *Tavy* is the practice of slash-and-burn agriculture; the land is cleared by cutting down the trees and burning the debris, and the cleared land is sown with crops (usually rice, but in this case maize). After a few years the soil is no longer productive, and a new field is needed to continue to cultivate the crops. Population pressure means that fields seldom, if ever, lay fallow for a sufficient length of time in order to regenerate nutrients, but more and more land must be cleared to produce enough maize, rice, and manioc to feed the population. Of course another danger related to *tavy* is wildfires, which are also common in this region and are also sometimes intentionally set by cattle rustlers to hide evidence of their crimes.

The conservation efforts at this site include raising environmental awareness and support for the community to control activities such shifting cultivation in their forest parcels.

To reduce the pressure on the forest, we have supported local people in the intensive production of maize and vegetables using footpumps to irrigate these crops and in animal husbandry (goat herding). In addition, three new community centers that include community libraries and classrooms have been constructed.

Anadabolava local partners:
Miramirasoa Analamaintso, Fitaovantsoa, Soanala, Agnalabo Mamokatra, Berivo Agnalasoa, Tafita, Tafavoaka, Mamelognarivo, Agnala Manirisoa, Tsilaisa, Komity Mpanorohevitra sy Mpanaramaso Fitantanana Ala (KMMFA) Anadabolava Betsimalaho

Left: A village outside the Anadabolava-Betsimalaho Thicket; top: Pristine thicket: Anadabolava at its best; bottom: A lone baobab bears testament to the former vegetation.

Analalava Forest

A partially degraded, low-elevation, humid forest, Analalava is located in eastern Madagascar close to the popular coastal tourist resort of Foulpointe and near the Ivoloina Zoo. The rich flora includes 26 palm species and several other plants endemic to the site and is the only remaining fragment of natural vegetation in the area. There are five species of lemur, including the nocturnal mouse lemur, and the Madagascar flying fox. The site was previously threatened by shifting cultivation, timber extraction, and burning by wildfires, but now is only threatened by wildfires. Without our intervention at this site, this forest, now regenerating strongly, would have disappeared.

Conservation efforts at this site include restoration of degraded parts of the forest using locally produced seedlings of native trees; protection of the forest from wildfires by means of an encircling firebreak; reforestation of the adjacent landscape with fast-growing alien trees to provide local people with alternative sources of fuel and timber; and raising environmental awareness by hosting local school children at the site for nature camps. The project also supports a program of research that aims to define the best protocols for forest restoration.

Because of its good potential as a tourist attraction, development at this site has included building overnight accommodations, a kitchen and dining facility, and classrooms for school groups. An observation tower for viewing fruit bats is under construction, and trail maintenance is ongoing. Water from the forest has been collected and used in a fish farm.

Left: Analalava Forest with one of the fish farm ponds; top: Chris Birkinshaw meets with local nurserymen at the nursery near Analalava; bottom: The entrance to the compound at Anlalava

Analalava local partner:
Velonala

Analavelona Sacred Forest

This unique sub-humid forest is located in the dry southwest and exists only because of a special local climate. A rare pocket of humidity creates a habitat for many species normally associated with the wetter eastern part of the country, and like Madagascar itself, its isolation has also led to the evolution of a number of locally endemic plants and animals. Fortunately, from the standpoint of conservation, the forest is largely untouched because the local Bara people consider it to be the home of the spirits of their ancestors, and therefore they strictly limit access and exploitation within this spiritual refuge. Among the approximately 260 plant species inventoried during the data collection around Analavelona Forest, around 100 species are collected exclusively from this forest, and more than 80 percent of them have spiritual or ritual uses. Recently these restrictions have begun to weaken as non-local people increasingly move into the area.

Our approach to conserving this forest is to support its traditional management and try to protect it from outside pressures. In addition, we support local communities in creating firebreaks to protect the site from wildfires that seem to occur with increasing frequency in the region.

Development focuses on new techniques for rice and maize cultivation as well as raising chickens and beans, and it also includes reforestation with fast-growing (but non-native) eucalyptus seedlings to provide firewood and timber.

Left: A majestic view of the sacred forest; top: Local villagers learn about plant identification; bottom: The nursery with protected seedlings outside the forest

Analavelona local partners:
Lovasoa, Soatanimbary Mandrosoa, Ampazotoa, Analavelo Miorisoa, Soamiranga Mitatanindraza, Vogniny Manantena, Maharitra, Alandraza Analavelona

Ankafobe Forest

This is a tiny forest fragment located in the Malagasy Highlands, a few hours' drive from Tana. Sub-humid evergreen forest patches dot the landscape within a number of adjacent valleys and are surrounded by poor, stubby grassland. The site is primarily important because it contains a large proportion of the world's total population of the tree *Schizolaena tampoketsana* (Sarcolaenaceae). As can be imagined on this high windswept landscape, the main threat to the Ankafobe Forest is burning from wildfires.

At Ankafobe, one of our main annual conservation activities is encircling the forest with double firebreak, as well as daily patrolling to detect wildfires. Without these measures the forest and its precious flora would likely be destroyed. In addition, each year 5,000 seedlings of locally produced native trees and shrubs are planted in parts of the forest that have already been degraded by fire in an effort to restore the forest.

Training and materials have been provided to allow local residents to produce seedlings of attractive native plants to sell to gardeners and landscape architects. The Ankafobe Forest is located adjacent to the main road between Tana and Mahajanga, and therefore has potential for tourism. This potential is being realized with the construction of trails, shelters, benches, and an interpretation center and boutique.

Left: Erosion of the plateau can be seen just behind the forest fragment at Ankafobe; top: Exercise in delimiting Ankafobe Forest community reserve with local stakeholders; bottom: Because the threat of fire is severe, local villagers must build and maintain a double firebreak

Ankafobe local partner:
Sohisika

Ankarabolava-Agnakatrika Forest

This highly fragmented, low-elevation humid forest is the only significant area of natural vegetation remaining in the Vangindrano District, near the southeastern coast of Madagascar. It has an exceptionally diverse flora, including many that are locally endemic. It is also a habitat for the critically endangered gray-headed lemur (*Eulemur cinereiceps*). In the last two decades, two-thirds of this forest has been destroyed by shifting cultivation and timber extraction. Although these threats have diminished over the years, shifting cultivation (*tavy*) continues to be practiced because of a lack of other land to farm, and timber extraction for construction is too high to be sustainable. More productive agriculture and new sources of timber and fuel are urgently required at this site.

Conservation efforts include creating forest corridors between currently isolated forest fragments, which are the natural habitat of the gray-headed lemur. Planting seedlings in the deforested areas between forest fragments should result in a more continuous forest, allowing the critically endangered lemur to remain genetically diverse rather than being split into numerous inbred sub-populations. This process involves supporting the local community in the implementation of rules for the sustainable exploitation of their timber sources; reorienting the local people away from shifting agriculture and toward more sustainable practices like permanent cultivation of crops such as cloves and coffee; and the installation of plantations of fast-growing alien trees as an alternative source of wood.

Development at Ankarabolava-Agnakatrika Forest includes providing tables and benches for 40 local public schools (previously the children had to sit on the dirt floor during classes), the construction of a local office for a management committee that is responsible for regulating the exploitation of timber, and installation of hundreds of signs marking the boundaries of the protected area.

Ankarabolva-Agnakatrika local partners:
Vohipaho Commune, Vemima, Mazavarano, Ami, Mangaresy, Ampitsahandaka, Miraindraiky, Valomars

Left: Ankarabolava-Agnakatrika Forest; Above: Sylvichadsia grandidieri *flowers at Ankarabolava-Agnakatrika*

Ibity Massif

The Ibity Massif is a spectacular 2,050-meter-high quartz mountain located in the Highlands south of Tana. Its natural vegetation includes gallery forest, sclerophyllous woodland, shrubby grassland, and rock vegetation. Of the 423 plant species that have been recorded from Ibity, 25 occur nowhere else, and most of the flora is not included in any other protected area. Our inventories suggest the site is less important for animals, although the site does support some threatened species of amphibians and reptiles and a rare highland colony of fruit bats. As is also true at Ankafobe, Ibity's biodiversity is threatened by wildfires, and although fires are a natural part of the ecology of the site, their actual high frequency inhibits the regeneration of the sclerophyllous woodland.

Conservation at this site centers on supporting a local management committee whose rangers patrol the site to detect and control threats (in particular wildfires but also illegal charcoal production and gold mining).

Local people are eager to conserve this site's natural riches because it is close to the major tourist route from central to southern Madagascar and therefore has huge potential to provide new livelihoods; in fact, plans have already been made to develop the area for tourism. We have assisted the local community by constructing a classroom for a local school, rehabilitating two local markets, and by providing support for a diversity of agricultural initiatives including the provision of ploughs and chickens.

*Left: Harmonious landscape surrounding Ibity Massif; top: Strange rock formations at Ibity Massif; bottom: Landscape at Ibity Massif with threatened Manambe palm (*Dypsis decipiens*)*

Ibity local partners:
Tamifa, Miavotra, Vandrikarana

Makirovana-Tsihomanaomby Forest

The fragmented, low-elevation, humid Makirovana-Tsihomanaomby Forest is situated on a range of small mountains in northeastern Madagascar. Our inventories of the area show many plant and animal species, including several locally endemic and threatened species. The forest also serves as an important water source for irrigating rice paddies on the extensive plains surrounding the mountains. This forest faces a number of threats: shifting cultivation (*tavy*), vanilla cultivation, and the extraction of timber and precious woods (most notably rosewood). This area has also seen a steep increase in immigrant and transient populations, so while the indigenous populations agree that forest conservation is critical, particularly because it is the source of water, the recent immigrants to the area are motivated differently and rely on unsustainable practices to gain their livelihoods because of a shortage of good farming land on the plains. Thus the main challenges to conserving this site are enforcing the *dina* with the non-local population and the illegal extraction of rosewood.

To promote development in the villages surrounding the proposed protected area we have funded the construction of two classrooms and a school library and supported an initiative to propagate and sell, through a specially constructed sales-point, seedlings of desirable and rare native trees (including rosewood) and crop plants such as cacao and coffee. Small parts of the forest degraded by shifting cultivation have been restored using locally propagated seedlings, but much more needs to be done.

Makirovana local partners:
Fandrefiala, Hevaniala, Hasiniala, Alamaintso, Alasoa, Alamamy, Alameva, Maitsolava, Mandroso, Loharanotsoa, Alakanto, Alatiavina, Alakitroka, Ravimaitso

Left: Makirovana-Tsihomanaomby; top right: Old primary school at Antanandava, a village next to Makirovana-Tsihommanaomby—we build a new one!; bottom right: Village nursery producing seedlings for forest restoration and for sale at Makirovana-Tsihomanaomby; This page: A local guide stands next to the biggest tree in the forest

Oronjia Forest

Oronjia is a fragment of dry deciduous forest located on a sand-covered limestone peninsula overlooking the huge natural harbor of Antsiranana on the northern tip of Madagascar. The major part of this protected area is within the boundaries of an almost-abandoned military base. The forest has been severely degraded because of timber extraction for construction and charcoal production, and parts of have been cleared for manioc cultivation. In spite of this extreme deforestation, our inventories suggest that the forest probably retains most of its original flora and, given the chance, could recover to its original condition.

Conservation efforts here principally revolve around slowing or halting deforestation by helping the local management committee implement a *dina* that controls the exploitation of natural resources; by establishing plantations of fast-growing alien trees adjacent to the reserve; and by facilitating access to alternative economic activities (such as vegetable growing, chicken raising, and handicraft production) for those who rely on charcoal production and shifting cultivation for their livelihoods. Parts of the forest that have been degraded though past activities are being restored with locally produced seedlings of native plants, and other parts of the forest that are designated for sustainable use are being enriched with useful native plants such as *Delonix velutina*, which is used to make canoes.

Sitting just on the gorgeous Diego Bay, this site has a huge potential for ecotourism. An ecotourism plan has already been conceived, and the phased implementation of this plan will soon begin. The aim is to maximize the benefits received by local people from this activity, and if this goal is achieved, we believe the future of Oronjia and its wonderful biodiversity will be secure.

Oronjia local partners:
Manovosoa, Fimpamaa, Mpamboly Miaradia, Tsimiroro, Riziky 2, Fimpamira, Malagasy Miray, Malagasy Tongasaina, Fikambanana Viavy Mpanao Massage, Tongasoa Miray.

Left: Delonix regia, *endemic to the dry forests in west and north Madagascar, is now grown all over the world in tropical countries as an ornamental tree.; top: Poverty is the biggest threat to conservation, and conserving the forest can often take second place to depleting its resources; bottom: These two crowned lemurs (*Eulemur coronatus*) are decidedly not camera-shy.*

Pointe à Larrée

This sandy peninsula juts out into the Indian Ocean opposite Île Ste Marie on Madagascar's east coast. The protected area represents a variety of vegetation types including littoral forest, low-elevation humid forest, swamps, and marshes, none of which are well represented in Madagascar's existing network of protected areas. The woody vegetation of the peninsula has been transformed over the past two decades because of exploitation of timber, shifting cultivation (*tavy*), and wildfires.

The main focus at this site is to support a management committee composed of local stakeholders in the implementation of a *dina* to control the exploitation of natural resources. We provide the committee with salaries and equipment for seven forest rangers who patrol the forest to detect infractions of the *dina*. Parts of the forest that are now badly degraded and susceptible to burning by wildfires have been protected with firebreaks.

In 2012, 80,000 seedlings of clove trees were produced and distributed to local farmers with the aim of providing them with an alternative livelihood and reorienting them from timber exploitation and agriculture based on shifting cultivation.

*Left: The forest abuts a deserted 8 km beach on the Indian Ocean; top: Madagascar Scops Owl (*Otus rutilus*) surprised during botanical inventory at Pointe à Larrée; bottom: Forest rangers responsible for detecting and reporting infractions to the* dina

Pointe à Larrée local partners:
Coopérative Samy Antsika, Lovasoa

Vohibe Forest

This low-elevation forest is located on the lower slopes of Madagascar's great eastern escarpment and is part of the Ankeniheny-Zahamena Forest Corridor. Thought to be important in enabling the flora of eastern Madagascar to shift their ranges in response to climate change, Vohibe is a rare example of almost pristine low-elevation forest. This vegetation type has the highest biological diversity of any ecosystem in Madagascar, yet it remains underrepresented in the country's network of protected areas. Vohibe has not been well explored because of its inaccessibility; nevertheless, to date, in collaboration with zoologists from Park Ivoloina, we have recorded 350 species of plants, 11 species of lemurs (including *Indri indri*), 80 species of birds, 23 species of reptiles, and 47 species of amphibians. Many more species await discovery. Human population density is low in this area, and consequently the Vohibe Forest is less threatened than other sites. However, there are occasional incidents of shifting cultivation and timber exploitation within the protected area, and with internal migration being an increasingly important issue in Madagascar, Vohibe's tranquil nature could quickly change.

Our activities at this site include supporting a local management committee that controls the exploitation of natural resources; the provision of alternative sources of timber through the plantation of seedlings of fast-growing alien species on land outside the forest; and the provision of training and materials to enable the adoption of new farming methods. At this site we have also invested heavily in environmental education using a range of methods including an annual biodiversity festival, video screenings, educational signs, and the creation of two "green clubs."

Development efforts here include animal husbandry (pigs and ducks); fish farming; vegetable, fruit, and coffee cultivation; a nursery; and improved rice cultivation. In addition, sewing machines have been provided so women have access a new source of income from the production of clothes.

Vohibe local partners:
Ambalabe Commune, Soavinala

Left: The waterfall is one of the most beautiful features of the Vohibe Forest; top: Community meeting to discuss useful plants, Ambalabe, outside of Vohibe Forest; bottom: A view of the cascades above the waterfall

This page left: A new species of Schefflera *(a member of the ivy and ginseng family, Araliaceae) discovered in 2006 by Pete Lowry in the Vohimena mountains of far southeastern Madagascar.; top right: A colorful hawkmoth caterpillar (family Sphingidae) stands out on a tree branch; middle right: Biodiversity awareness celebration in Pointe à Larrée; bottom right: Vigna collected during inventory of inselberg plants near Andranomena*

Opposite page: top left: Madagascar at its most dramatic: the gorges of Makay; bottom left: Collecting different leaf shapes to teach diversity to primary school children at the Ankafobe Forest; top right: Many of our conservations sites are valued as a source of water for the irrigation of rice fields; bottom right: The impact of shifting cultivation on the Ankarabolava-Agnakatrika Forest

25 YEARS

from plant collecting to a multi-faceted research, training, and conservation program

| 1970 | 1975 | 1980 | 1985 | 1990 |

*First visited by
MBG staff in 1972*

*Resident botanist
starting in 1984*

*MBG office established
in May 1987*

MBG's MADAGASCAR PROGRAM MAJOR LANDMARKS

Our Madagascar program is most remarkable because of its multi-faceted nature. It differs from other Garden programs in how it has evolved beyond our typical research activities—collecting plants, identifying them, and focusing on taxonomy. Over the last 25 years, we have progressed from an exclusive focus on exploration and taxonomic research to a program that includes training, infrastructure support, and collaborative conservation with local communities living ajacent to protected areas (parks and reserves). Each facet of what we do speaks simultaneously to the Garden's mission and to Madagascar's needs.

Opposite page, top: (left to right) Mr. Léon Rajaobelina, Ambassador of Madagascar in the U.S.; Dr. Porter P. Lowry II; Dr. Ignace Rakoto, Minister of Higher Education and Scientific Research; Dr. Voara Randrianasolo, Director of PBZT; Dr. Enrique Forrero, Former Director of Research at MBG; and Dr. Peter Raven, President Emeritus of MBG during the signing ceremony of our first ever collaboration agreement protocol with the Parc Botanique et Zoologique de Tsimbazaza(PBZT) in St. Louis.

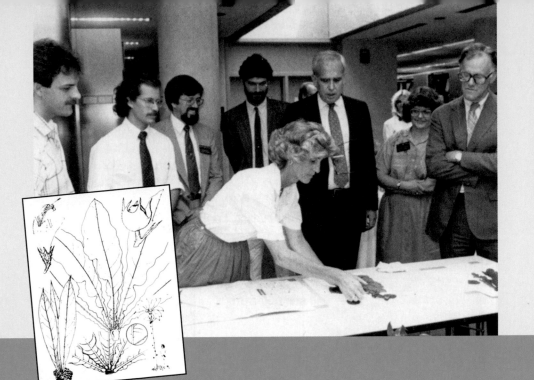

(Left to right): Jon Ricketson, former collections manager; George E. Schatz, curator; Enrique Forero, former director of research; Porter P. Lowry II, Curator and Head of the Africa and Madagascar Department; Peter H. Raven, President Emeritus; Nancy R. Morin, former curator of the herbarium; and John H. Biggs, former head of the Board of Trustees look on as Gigi Hill, former plant mounter, prepares the specimen of Vitex masoalensis G. E Schatz, marking the 3.5 millionth specimen collected in the herbarium in St. Louis, 1988; inset: Drawing by John Myers.

1995   2000   2005   2010   2015

Formal recognition as a locally operating international organization in 1993

Strategic decision to formally engage in community-based conservation efforts in 2002

Opposite page, bottom: The house which served as the first office and residence for the local staff. It is still used as an office and temporary lodging for visiting staff and other researchers.

Center: Jim Miller, former Director of the William L. Brown Center in St. Louis, and local guides on the summit of Marojejy (2132 m), one of Madagascar's highest mountains, February, 1989; inset: Page one of the paper that kick started the Garden's research and conservation program in Madagascar.

Clockwise from right: Armand Randrianasolo stands next to a successfully growing tree as part of a reforestation effort in Vohibe Forest; George Schatz (left) visits the newly renovated greenhouse at PBZT with then Director Voara Randrianasolo (right), 1989; Pete Phillipson and Charles Rakotovao press specimens of Xerophyta nandrasanae, *another new species known from a single site on Madagascar's central plateau, 2003; Pete Lowry pressing a specimen of* Xerophyta schatzii, *a new species collected near Andohahela national park in southwestern Madagascar, 2008; We would not be the successful program we are without our field collectors and field botanists*

HISTORY

A 1974 paper co-authored by Missouri Botanical Garden President Emeritus Peter Raven and paleontologist Daniel Axelrod (Raven, P.H., Axelrod, D.I. (1974) Angiosperm biogeography and past continental movements. *Ann. Mo. Bot. Gard.* 61: 539-673) became the starting point of the Garden's scientific involvement in Madagascar. Dr. Raven had identified Madagascar as a key place for understanding the evolution of flowering plants, and designated the island as a "biodiversity hot spot" even before that term was coined.

At about the same time this paper was published, the Garden received a W. Alton Jones Foundation grant just as Madagascar was opening up to "western" countries. The new grant supported the salaries of two researchers (Pete Lowry, who was initially based in St. Louis but shortly took up residence at the Paris herbarium, and George Schatz, who was initially posted in Madagascar) plus one technical staff member in St. Louis. The Garden's involvement now constituted a "program." After this, the Garden's program in Madagascar grew rapidly as we identified new ways in which we could carry out interesting and relevant work.

Left: Jean François (from Park Ivoloina) extracting seeds for cultivation as part of an ex situ plant conservation project; right: Community meeting with Armand and two Peace Corps volunteers

The early days of the program focused on discovery and documentation of the rich and diverse flora that makes Madagascar an important research site. Garden researchers made expeditions to many of the "classic" botanical localities, primarily in the eastern part of the country, where most of the remaining forest exists. We established collaborative partnerships with the herbarium staff at Tsimbazaza Botanical and Zoological Park (known in Madagascar as Parc Botanique et Zoologique de Tsimbazaza or PBZT) and other local institutions. These partnerships were undertaken to ensure the longevity of the program by establishing and nurturing local scientific ties. With a researcher based in Tana, and collaborative relationships with other local institutions, the next step was to begin

actively training local Malagasy botanists who would soon become the foundation of the Madagascar program.

Over the next decade, exploration, discovery, and documentation took the staff into areas that had not been explored before, including desert and highland areas in central and southern Madagascar. Some botanists who had been trained in Madagascar joined the staff and, with the help of Garden-trained field collectors, were actively compiling inventories and pursuing taxonomic research on plant groups found in many of these areas. We became active in publishing new species, revising genera, and publishing practical guidebooks. During this time, the TROPICOS-based Catalogue of the Vascular Plants of Madagascar was first developed. Although threats to the local biodiversity were evident, and Madagascar had been officially designated one of the world's most endangered "hot spots," conservation efforts were indirect: we provided data and collaborated with conservation projects by making botanical expertise available. Taxonomic studies and field work remained the primary focus of the program.

A strategic retreat in 2002 shifted that focus. Staff members, including several from Madagascar, convened in St. Louis to discuss the future of the program. During the retreat, the Malagasy staff expressed a strong desire for the Garden to become engaged in *doing* conservation, not just conducting research and generating information that could be used for conservation. Decisions made during the retreat prompted a study that identified 79 sites around the island in need of protection, collectively known as Priority Areas for Plant Conservation (PAPCs), which the staff in Tana lobbied to have incorporated into the national protected areas network. We selected 11 of these areas (described in this book) from among the 79, chosen because of their exceptionally rich floras and their extremely endangered status, but of potentially limited interest to other conservation organizations. Our program now includes community-based conservation planning, capacity building, education, and advocacy, in addition to taxonomic research and discovery.

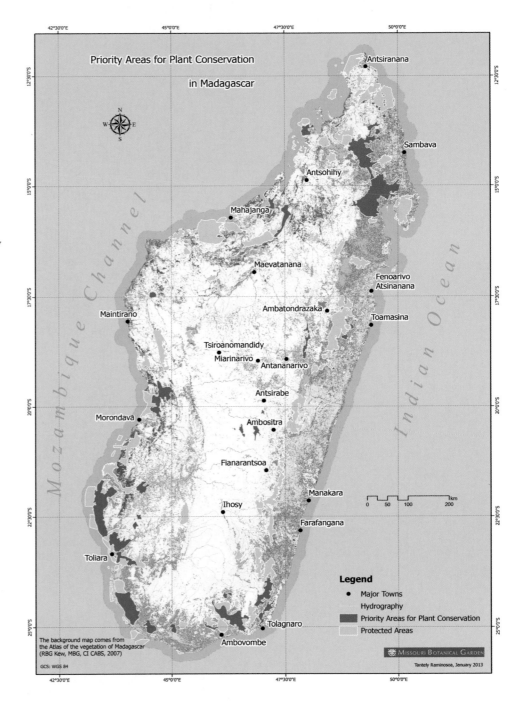

This page: top left: Staff photo during a visit to Madagascar by Garden president Dr. Peter Wyse Jackson; middle left: The staff gathers daily for a family-style meal at lunch; bottom: Group photo during a staff outing to Ankafobe Forest; right: Patrice in Vohibe

Opposite page top: Garden staff, students, and associates made up the largest delegation attending the 18th Congress of the Association for the Taxonomic Study of the Tropical African Flora, held for the first time ever in Madagascar, in May 2010; middle left: Field botanists preparing plant samples for transport to Tana; middle right: Sylvie Andriambololonera, Marina Rabarimanarivo, and Hans Rajaonera at PBZT in Tana; bottom: Madagascar staff in 2002.

STAFF

We have over 50 full-time staff members in Madagascar and another 100 or so contractual staff members around the country. Our local staff in Tana manages the program and therefore has a level of ownership in it that is truly unique among research and conservation organizations in Madagascar. Several researchers, including Pete Lowry, Armand Randrianasolo, Pete Phillipson, and George Schatz travel to Madagascar two or three times a year, and other Garden staff cycle through periodically. In addition, some of the Madagascar-based staff also travels to the United States, France, and occasionally elsewhere for their work. The Garden has an *Accord de Siège* which provides us semi-diplomatic status in recognition of the fact that our work makes a positive contribution to the country.

EDUCATION

Over the last 25 years, the Garden has been involved in providing educational opportunities in botany and conservation for nearly 100 Malagasy nationals. Three people have pursued advanced degrees at the University of Missouri, St. Louis, while most of the others have been involved in master's level work at the University of Antanarivo in the capital city. Typically after a student has completed his or her undergraduate degree, we sponsor and supervise their master's work, allowing them do a field-based project that is appropriate for a thesis, but will also contribute to our own research and conservation work. Over the years, because we have been able to offer compelling projects and provide good mentorship, the universities have sent their best students to us. Nearly all of our current staff in Madagascar came through this training program, which gave them a good background for working with us while providing us an opportunity to see them in action and to select the most promising students to join our team.

We also help others succeed in their work by assisting visiting scientists. In addition to advising them on how to conduct their work responsibly in Madagascar, we help them apply for research permits, secure transportation and other in-country assistance, and offer them reduced-rate housing in Tana. Some of these scientists come to work independently of our program, while others visit with the intention of starting a collaborative relationship with the Garden.

Top left: Rondro Ramananjanahary and Richard Razakamalala work together to press plants; middle left: Nivo Rakotoarivelo and Tabita Randrinarivony; bottom left: Elise Buisson, a collaborator from the University of Avignon, France, measuring seedling after a restoration work at Andohahela corridor/Fort Dauphin; top right: Vohitsanavo Massif, in the front of Anja Park; bottom right: Students from GRENE University of Toamasina (Tamatave) helping us to reinforce the wild population of the critically endangered tree Sarcolaena grandiflora.

THE NEXT 25 YEARS

During the next quarter-century, one of the biggest challenges in Madagascar will be how the country deals with climate change. How will climate change affect the native flora? Studies done by our staff suggest that while some species will be able to adapt to changes and other plants will migrate to new areas where they can grow, many species will no longer be able to survive where they are found today and won't be able to reach suitable areas since the native vegetation in most parts of the country is no longer continuous. In order to survive, many, if not all, of these plants will require intervention by humans, moving and planting them in places where they can grow and prosper. Madagascar has a rapidly growing, extremely poor population. Climate change will adversely affect agriculture as well and will reduce yields of many crops in areas where they already barely feed today's population, increasing pressure on natural resources.

Our hope is that we can secure our 11 conservation sites and pass conservation and development efforts entirely into the hands of their surrounding communities. We would like to explore the island's remaining botanical frontiers and to complete the Madagascar Catalogue project. We plan to continue to train promising young scientists and help them secure meaningful employment. The Madagascar Program can serve as a model for expanding the Garden's work in other parts of the world. The program incorporates a diverse set of activities in research and conservation, and it has successfully empowered our local staff members, who are taking our core work (which has historically focused on discovery and taxonomic research) and leveraging it to do a whole new set of good things—for Madagascar and for the planet. We hope to continue the work of the Garden's Madagascar program for many decades to come.

Top left: Central highlands near Ankafobe; middle left: Corn and garlic dry outside a food storage hut in Analavelona; bottom left: Fish farm near Anlalava; top right: Tree fern Cyathea sp. in Vohibe; bottom right: Students at Antanandava school

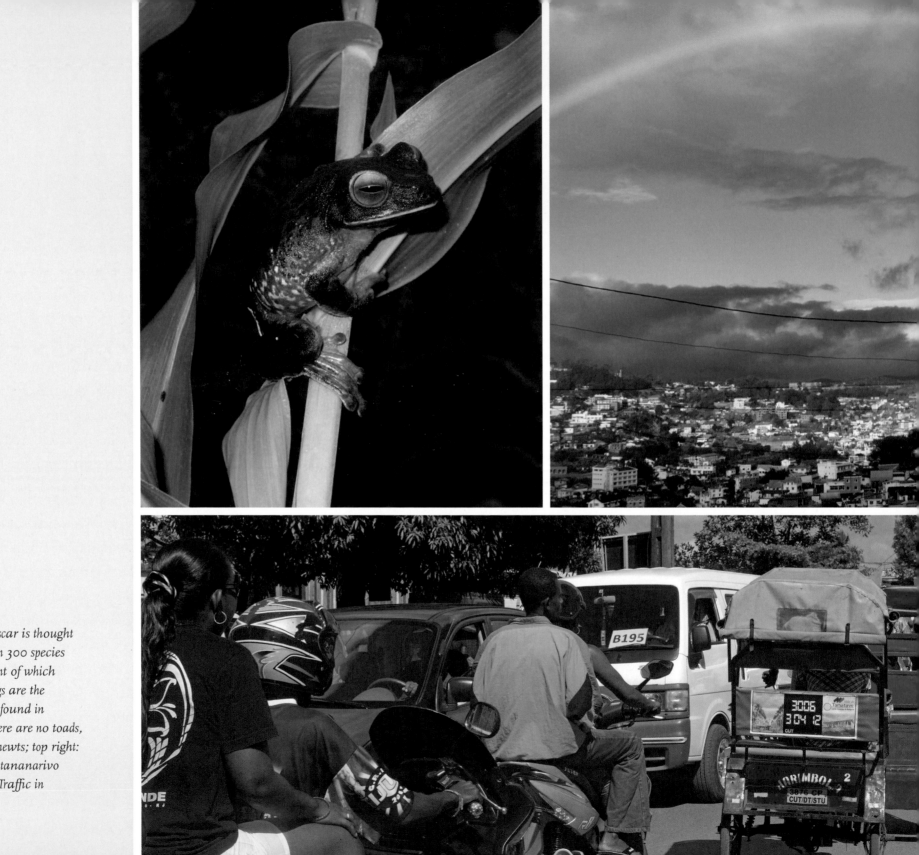

Top left: Madagascar is thought to have more than 300 species of frogs, 99 percent of which are endemic. Frogs are the only amphibians found in Madagascar—there are no toads, salamanders, or newts; top right: Rainbow over Antananarivo (Tana); bottom: Traffic in Toamasina

The Missouri Botanical Garden in Madagascar wishes to thank the following for their generous support and collaboration.

African Plants Initiative (API)
Ambatovy S.A.
Andrew W. Mellon Foundation
Asity Madagascar
Arboretum Antsokay
Aronson Charitable Trust
Association du Réseau des Systèmes d'Information Environnementale (ARSIE)
Beneficia Foundation
Bergius Foundation
Biodiversity Conservation Madagascar (BCM)
Blessing Basket Project (BBP)
Botanic Gardens Conservation International (BGCI)
California Academy of Science (CAS)
CARE International
Carnegie Institution
Catholic Relief Services (CRS)
Centre National d'Application des Recherches Pharmaceutiques (CNARP)
Centre National de Recherche Océanographiques (CNRO)
Centre National de Recherche sur l'Environnement (CNRE)
Centre VALBIO
Comité Français de l'Union Internationale pour la Conservation de la Nature (CF-UICN)
Conservation International Madagascar (CI Mad)
Conservatoire et Jardin Botaniques – Ville de Genève
Consulate of Monaco in Madagascar
Critical Ecosystem Partnership Fund (CEPF)
Département Biologie et Ecologie Végétales (DBEV), Faculté des Sciences, Université d'Antananarivo
Direction Générale des Forêts (Ministère de l'Environnement et des Forêts)
Dr. James and Janelle Evans Criscione
Duke Lemur Center
Durrell Wildlife Conservation Trust (DWCT)
Ecole Supérieure des Sciences Agronomiques/ Forêt, Université d'Antananarivo (ESSA)
European Association of Zoos and Aquaria (EAZA)
Fanamby NGO
Fauna and Flora International (FFI)

FOFIFA
Fondation pour les Aires Protégées et la Biodiversité de Madagascar (FAPBM)
Fonds Français pour l'Environnement Mondial (FFEM)
GEF Small Grants Programme
Gestion de Ressources Naturelles et Environnement, Université de Toamasina (GRENE)
Global Colors Foundation
Helmsley Charitable Trust
HOLCIM S.A.Madagascar
Hotel Eulophiella
Idaho Botanical Research Foundation
Idea Wild
Institut Malgache de la Recherche Appliquée (IMRA)
Institute for the Conservation of Tropical Environments (ICTE)
International Cooperative Biodiversity Group (ICBG)
International Foundation
IUCN Madagascar Plant Specialist Group
JRS Foundation
JSTOR
Lemur Conservation Foundation
Liz Claiborne and Art Ortenberg Foundation (LCAOF)
MacArthur Foundation
Madagascar Conservation & Development (MCD)
Madagascar Fauna Group (MFG)
Madagascar Institut pour la Conservation des Ecosystèmes Tropicaux (MICET)
Madagascar National Parks (MNP)
Madagasikara Voakajy NGO
Man And The Environment (MATE)
Marisla Foundation
Ministère de l'Education Nationale – Madagascar
Ministère des Affaires Etrangères – Madagascar
Mohamed bin Zayed Species Conservation Fund (MBZ)
Muséum National d'Histoire Naturelle (MNHN) de Paris, France
National Cancer Institute (NCI)
National Geographic Society (NGS)
National Institutes of Health (NIH)
National Science Foundation (NSF)
Naturevolution (Makay)

Observatoire National de l'Environnement et du Secteur Forestier (ONESF)
Office National pour l'Environnement (ONE)
Parc Botanique et Zoologique de Tsimbazaza (PBZT)
Phyto-logic
Population Services International (PSI)
Private Agencies Collaborating Together (PACT)
QMM/Rio Tinto S.A
Rainforest Alliance (RA)
Reggio Terzo Mondo
Réseau des Educateurs et Professionnels de la Conservation (REPC)
Restoration of Natural Capital Alliance
Rhodes University
Richard & Rhoda Goldman Fund
Royal Botanic Gardens, Kew (RBG, Kew)
Seacology
Service d'Appui à la Gestion Environnementale (SAGE)
Silo National des Graines Forestières
Stony Brook University
Sud Expert Plantes (Ministères français des Affaires étrangères et européennes)
Tany Meva Foundation
Tuléar Sands Project
United States Agency for International Development (USAID)
Université Nord Antsiranana (UNA)
University of Fianarantsoa
University of Maryland Center for Environmental Studies
University of Missouri – St. Louis (UMSL)
University of Rostock
US Embassy in Madagascar
US Peace Corps
Vahatra
Virginia Polytechnic Institute and State University
W. Alton Jones Foundation
Washington University in St. Louis, Missouri (WUSL)
Welt Hunger Hilfe
Whitley Fund for Nature
Wildlife Conservation Society (WCS)
World Wildlife Fund (WWF)
Zoo Zürich